Children, Nature and Cities

That children need nature for health and well-being is widely accepted, but what type of nature? Specifically, what type of nature is not only necessary but realistically available in the complex and rapidly changing worlds that children currently live in?

This book examines child-nature definitions through two related concepts: the need for connecting to nature and the processes by which opportunities for such contact can be enhanced. It analyses the available nature from a scientific perspective of habitats, species and environments, together with the role of planning, to identify how children in cities can and do connect with nature. This book challenges the notion of a universal child and childhood by recognizing children's diverse life worlds and experiences which guide them into different and complex ways of interacting with the natural world.

Unfortunately not all children have the freedom to access the nature that is present in the cities where they live. This book addresses the challenge of designing biodiverse cities in which nature is readily accessible to children.

Claire Freeman is Professor in Geography at the University of Otago, Dunedin, New Zealand, where she teaches Master of Planning and Geography students. Her research is in environmental planning, where she works on planning with children, urban nature and community planning. She has previously worked at universities in South Africa and the UK and as a planner with the Urban Wildlife Trust, UK.

Yolanda van Heezik is Associate Professor in the Zoology Department at the University of Otago, where she teaches wildlife management and conservation biology. Her research has focused on the spatial ecology of urban wildlife, biodiversity of private gardens, residents' attitudes, motivations and knowledge about biodiversity and penguin ecology. She has previously worked in The Netherlands, South Africa and Saudi Arabia.

Routledge Spaces of Childhood and Youth Series

Edited by Peter Kraftl and John Horton

The *Routledge Spaces of Childhood and Youth Series* provides a forum for original, interdisciplinary and cutting edge research to explore the lives of children and young people across the social sciences and humanities. Reflecting contemporary interest in spatial processes and metaphors across several disciplines, titles within the series explore a range of ways in which concepts such as space, place, spatiality, geographical scale, movement/mobilities, networks and flows may be deployed in childhood and youth scholarship. This series provides a forum for new theoretical, empirical and methodological perspectives and ground-breaking research that reflects the wealth of research currently being undertaken. Proposals that are cross-disciplinary, comparative and/or use mixed or creative methods are particularly welcomed, as are proposals that offer critical perspectives on the role of spatial theory in understanding children and young people's lives. The series is aimed at upper-level undergraduates, research students and academics, appealing to geographers as well as the broader social sciences, arts and humanities.

Young Entrepreneurs in Sub-Saharan Africa
Edited by Katherine Gough and Thilde Langevang

Children, Young People and Care
Edited by John Horton and Michelle Pyer

Youth Activism and Solidarity
The Non-Stop Picket Against Apartheid
Gavin Brown and Helen Yaffe

Children, Nature and Cities
Rethinking the Connections
Claire Freeman and Yolanda van Heezik

For more information about this series, please visit: www.routledge.com/Routledge-Spaces-of-Childhood-and-Youth-Series/book-series/RSCYS

Children, Nature and Cities

Rethinking the Connections

Claire Freeman and Yolanda van Heezik

Routledge
Taylor & Francis Group
LONDON AND NEW YORK

First published 2018 by Routledge

2 Park Square, Milton Park, Abingdon, Oxfordshire OX14 4RN
52 Vanderbilt Avenue, New York, NY 10017

Routledge is an imprint of the Taylor & Francis Group, an informa business

First issued in paperback 2020

British Library Cataloguing-in-Publication Data
A catalogue record for this book is available from the British Library.

Library of Congress Cataloging-in-Publication Data
A catalog record for this book has been requested

ISBN: 978-1-138-94255-4 (hbk)
ISBN: 978-0-367-58943-1 (pbk)

Typeset in Times New Roman
by Apex CoVantage, LLC

Contents

Figures

Tables

About the authors

Claire Freeman and Yolanda van Heezik have worked together for around fifteen years on projects that look at relationships between people and the natural environment in cities. Their joint projects include urban habitat mapping, garden biodiversity, city children's experience of nature and older people's connections with nature. They are currently working on a multi-city study exploring how city dwellers value and use green space and motivations behind engagement in biodiversity conservation. Both are involved in a community project to try to bring the kaka, an endangered New Zealand parrot, back into their city.

Foreword

Paul Tranter

Several years ago, I read a draft chapter that Claire Freeman had written for a book I was co-authoring with her: *Children and their Urban Environment: Changing Worlds*. The chapter was titled "Natural Space". I remember being so deeply inspired by the arguments in this chapter that I rang my wife and arranged to meet her near Lake Burley Griffin. We walked through a forest and experienced some of the therapeutic contact with nature that Claire had explained in her chapter. I believe that readers of *Children, Nature and Cites* will be similarly prompted into action, equipped with how-to guidelines for enabling the greening of cities and promoting children's contact with nature.

For an increasing majority of the world's children, learning about nature and having contact with nature occurs mainly in cities. The benefits of contact with nature are immense; when children have increased contact with nature, everyone benefits. Yet, as Claire Freeman and Yolanda van Heezik explain, despite the wealth of evidence about the health, developmental and happiness benefits of contact with nature, other evidence points to declines in both children's preference for natural spaces and their access to nature over the last decade. In a time-pressured world, children are increasingly likely to be occupied in adult-organized, indoor or virtual activities. While children have the capacity and the motivation to seek out nature if it is present, there are increasing constraints (particularly societal constraints) on children's freedom to do so.

Why is nature so important for children? How has children's opportunity for contact with nature changed? How is this different in different nations, cities and suburbs? What is nature? How can children's contact with nature be enhanced? These and many others are the questions that this book considers. To answer these questions requires the understanding of a multitude of interrelated topics, including urban habitats and biodiversity, density, socio-economic status, ethnicity, traffic, pet ownership, social connection, parental fears, international differences and many others.

Claire Freeman and Yolanda van Heezik are ideally placed to handle this complexity. They have worked together on their shared passions of children and nature for several years, including on a project that informs some of the discussion in this book, the *Natural Neighbourhoods for City Children* project. Importantly, both authors are parents, and the book includes insights from their children's own experience, for example, adapting to different spaces for contact with nature when moving from the UK to New Zealand. Claire's research expertise in planning with children and urban nature meld well with Yolanda's expertise in conservation biology and urban wildlife, providing the necessary multi-disciplinary perspective required to appreciate that "good habitat for wildlife is also good habitat for children".

Freeman and van Heezik carefully seek out the views of children, for example, using rich interpretations of children's drawings of their contact with nature. Even in neighbourhoods that governments and media reports see as deprived or depressing, children see the positives of their environment and often have strong feelings of attachment and in some cases greater freedom and hence independent access to nature than children in more advantaged neighbourhoods. Of course, in some disadvantaged areas, children's access to nature can be restricted, for example by fears of violence. Using a careful analysis of case studies, this book demonstrates how to re-introduce nature to where children live: their homes, schools, streets, hospitals, gardens and city-farms. In explaining what is needed to create more biodiverse cities for children, the book considers relevant professionals, agencies and organisations, as well as children themselves, both as users of nature and as researchers and contributors to the creation of more natural environments.

The views of children inspire a rethink about what nature means and how children can engage with nature. With appropriate attention to design and careful involvement of all citizens, especially children, even the most challenging environments can offer nature-rich experiences. Exciting international examples demonstrate how city biodiversity can be expanded so that children can experience nature in their daily lives.

The convincing arguments are brought even more alive by dozens of carefully chosen photographs, many illustrating successful (and less successful) examples of initiatives to enhance children's contact with nature. From snakes to geckos, gardens to forests and barren streets to rehabilitated river frontages, the truly international coverage of the book is reflected in the diverse places where these photographs were taken, including London and other English cities, Belfast, Amsterdam, Stockholm, Bratislava, Punta del Este, East London (South Africa), Kathmandu, Singapore, Denpasar, Seoul, Thiruvananthapuram, Mumbai, Hong Kong, Suva, Samoa, and several cities in Australia and New Zealand. These photographs provide encouraging evidence that it is possible to enhance children's connection with nature.

Like Colin Ward in his book *The Child in the City*, Freeman and van Heezik take a 'whole of city' approach, enabling contact with nature as part of children's everyday lives in their homes, schools, streets and local neighbourhoods, and the left-over spaces in cities, which, though often undervalued, can provide some of the richest natural areas in a city. Freiburg, Germany, is presented as one exemplar of forward thinking and an integrated vision for all elements of the city (e.g., transport, housing, playspaces, energy and waste) that enhance urban life. In addition to detailed insights into the ecology of cities, Freeman and van Heezik demonstrate an appreciation of the value of such simple, pleasurable contact with nature as feeding ducks, or incidental contact with nature within the house, including spiders, moths and lizards.

Children, Nature and Cities is written at a time of immense global challenges. As well as global climate change, loss of ecological biodiversity and human population pressures, our global civilisation is transforming cities at an unprecedented rate. These challenges can seem immense and overwhelming. People increasingly need an injection of hope and optimism about the future. This book provides this injection along with a set of strategies to turn this hope into reality. The positive stance taken in this book reflects the positive views of children about the nature present in their cities, and the voices of children provide a pivotal part of the book. *Children, Nature and Cities* will leave readers more optimistic about the future well-being of children and the health of the planet.

Paul Tranter is a geographer in the School of Physical, Environmental and Mathematical Sciences at UNSW Canberra. His research on children's geographies began in the early 1990s, with the first Australian and New Zealand study of children's independent mobility. Since then, Paul has researched children's active travel, child health and well-being, play and risk, environmental learning, child-friendly environments and children's rights. His research is based on the policy objective of creating better environments for all citizens.

Acknowledgements

We would like to acknowledge the substantial contributions made by a number of people who made this book possible. Firstly all the children and teachers in classes 5 and 6 from the schools in Auckland, Dunedin and Wellington who contributed their time and expertise to the Natural Neighbourhoods for City Children project. Our thanks also to the children from Dunedin North Intermediate and North East Valley Normal Primary schools who contributed the quotes and questions that we used in this book, and their teachers, who enthusiastically allowed us access to their classes. To the many children and families who agreed to us using their photos. Chris Garden of the Geography Department who produced some of the figures, Aviva Stein who worked with the children, and Kate Hand who undertook the biodiversity surveys and analysis. Yvette Buttery who proof-read the book and provided general support. Finally, the Royal Society of New Zealand for the Marsden Fund Grant U001222 that made the Natural Neighbourhoods for City Children study possible. To all of you we say thank you.

Part I

Children and nature – approaches to nature

1 Conceptions of nature, contested definitions and approaches

What type of nature is needed and why?

That children need nature for health and well-being is widely accepted but what type of nature? Specifically, what type of nature is not only necessary but also realistically available in the complex and rapidly changing urban worlds within which children currently live? Much recent interest in children's connection with nature has come about because of growing concern around children's perceived lack of nature connection. In this book, we explore children's experiences and connection to nature. We examine different ways to approach and analyse the nature relationship, different ways of experiencing nature and of supporting and enhancing nature in the city. Our particular focus is on the relationship children have with nature and natural places in the city. In addition, we also adopt an ecological perspective when examining the biodiversity that both is and could be present in the urban environment and explore what this tells us about the potential for better engagement with nature.

In this book we examine child-nature definitions, conceptions and relationships. We recognise that these relationships can be 'messy', contradictory, complex and energising in real world experiential contexts. This approach is preferred over the more usual reference to romanticised notions of untouched or unpeopled nature and rural idylls. We explore two related concepts, ways of connecting to nature and the processes by which opportunities for such contact can be enhanced. We adopt a positive and optimistic view in that we have, as our core theme, a belief that connections with nature for children today are not only desirable but are possible in nearly all urban contexts. This is not to deny the importance of experiencing the wild and pristine but rather to emphasise and celebrate the common, the accessible and the interactive. In cities, if the opportunity is there, nature will use it to colonise the available space, and if nature is present, children can seek it out.

One example of a coloniser is the common mallard (*Anas platyrhynchos*), a duck that even though it has always been widely distributed throughout the temperate and subtropical Americas, Europe, Asia and North Africa has continued to spread its wings. It is now the commonest duck in our home town of Dunedin. Like many urban residents, it adapts well to a variety of living conditions. It can live in quite large population groups and is not too fussy about what it eats or with whom or even with what other type of ducks it mates, leading sometimes to a quite 'crossbred' urban population.[1] Just as these ducks have sought out human environments, children seek out ducks. Wherever ducks gather, so too will children for whom ducks provide one of the most enduring of nature interactions: enduring because these children will in turn bring their own grandchildren for the time-honoured ritual of 'feeding the ducks'. Yet, this common and almost universal practice, which is central to connecting people and nature, goes unrecognised by most nature advocates.

Why is it that so little weight tends to be given to those common and often 'fun' interactions, such as duck feeding, messing around in the mud, tree climbing, fishing with a net or rolling on the grass? Is it because these and other childhood activities are seen as ordinary, everyday and not worthy of serious scientific attention? Whatever the reason, we argue that the everyday does matter and the ways in which nature makes its presence felt in childhood matters, as it is here in childhood that, we hope, the seeds of a lifelong relationship with nature are sown.

There is no universal child. Children live in and experience different life worlds, which, in turn, guide them into different and complex ways of interacting or not interacting with the natural world. Geographic location, culture, societal characteristics, race and ethnicity, religion, age, socio-economic status, education and a host of other influences determine these life worlds. These influences can be present at a macro level and expressed in a socio-cultural attitude to nature as in a predominantly Buddhist society or at the very micro level, where the poor physical health of a parent may limit the family's access to walks in the natural spaces present in their neighbourhood. While some experiences may be generally available and characteristic of children in a particular place or cohort, individual characteristics, such as a disability, can and do create differential access to nature connection experiences. These differences mean that in any nature-enhancement initiative, professionals need to be fully informed of the factors in society that act at the general and the individual child level to enable or frustrate nature connection.

This book derives in part from the *Natural Neighbourhoods for City Children* study, funded by the Royal Society of New Zealand Marsden Fund (Freeman et al., 2015). Where appropriate, we include and reflect on findings from this study. The study was undertaken with 187 children in nine neighbourhoods in three different cities and included children from diverse socio-economic and ethnic backgrounds. The interactions we had with the children and their teachers acted as the catalyst for this book, encouraging us to think more widely on the role of nature in city children's lives. By most standards, New Zealand cities are green and biodiverse, although much of the flora and fauna in cities is not of New Zealand origin but made up of introduced exotic species that do not always sit well within native ecosystems. Rats, possums, hedgehogs, house sparrows, roses, loquats, lawns, bumblebees and the mallard are part of a seemingly endless list of introduced species resident in the cities in which children live. There are native species too but not always as prevalent or recognised. This complex exotic, indigenous biodiverse mix and the values and experiences associated with it also form part of this book as we consider what nature means in this increasingly globalised 'natural' world in which children grow up.

In our *Natural Neighbourhoods for City Children* study, we were intrigued by the absence of a value-laden approach of our children to nature. They liked or disliked nature, for example, the ubiquitous rat or trees in the street, on their own merits, rather than according to wider societal or scientific views on whether the species is native or alien or the aesthetic values placed on particular species. Our children were almost universally positive in their views on the nature present in their cities. This was a view present regardless of whether they lived in state-provided apartments, in low- or high-income housing estates or in inner-city or suburban housing. They all relished the opportunity to talk about the nature that they saw and encountered in their everyday lives. There were differences: lower-income children generally had less access to the more biodiverse environments (mainly private gardens) in their neighbourhoods compared to children in higher-income neighbourhoods. Nevertheless, even there, they showed an appreciation for what was there and did not see themselves or the places they lived as nature-deficient. We take our cue for this book from these children and attempt to focus on the positive opportunities present in our cities. Rather than being doom-and-gloom proponents, we look to how cities can and do work for children and to see if children can be supported in accessing the nature that is and could be there.

We traverse a wide global terrain for the book but recognise the focus is mostly on children and nature in western cities. In places, we use examples and case studies from New Zealand to illustrate the book. We recognise that the New Zealand experience is particular to these islands, but many of the challenges that New Zealand cities and New Zealand children face have resonance for wider global thinking on children's experiences. In some respects, the New Zealand experience is unusually insightful, and, in this, we look to the experiences of New Zealand's indigenous Māori people and New Zealand's position as part of the Pacific realm. Children's nature connection is informed by the specific geographical, historic and cultural contexts that shape their natural lives and by the globally challenging environmental context that will increasingly determine their future. We next turn our attention to this challenging context.

The environmental/nature challenge; cities and children under stress

In writing this book, we recognise and respond to a number of environmental challenges as they relate to urban living. These are threefold: the challenges nature itself faces globally that impact on cities at the macro level (e.g., population growth, climate change), the challenges that nature faces in the urban environment (e.g., habitat loss, fragmentation, species introductions), and the challenges children face in connecting to nature in these changing environments. These challenges occur at different geographic and temporal scales, but there is reasonable uniformity of agreement on what the global challenges are and the significant role of urban environments in these challenges. The United Kingdom's (UK's) Nature Conservancy (2016) identifies five key global challenges:

1 Protecting water
2 Action on climate change
3 Saving oceans
4 Conserving land
5 Transforming cities

The five problems, though presented as a list of five separate items, are deeply interconnected. For example, cities are impacted on by climate change with rising seas particularly affecting coastal settlements. As in many other parts of the world, New Zealand is one of the most affected countries because its major cities are coastally located. The quality of oceans is impacted on by urban economic development, its activities, waste and urban growth, which removes land available for wildlife and conservation. Cities are also associated with a number of key changes identified here that have relevance for children's nature interaction.

1 More people will live in cities

Within cities, the process of transformation identified as the fifth challenge by the Nature Conservancy Council (mentioned earlier) has a number of key elements. More children now live in cities than in rural areas. In 1800, only 3% of humanity lived in cities. The United Nations (UN) projects that 70% of people will live in cities by 2050 (World Wildlife Fund (WWF), 2012) with the proportion that are urban-based already being substantial in many countries. For example, in New Zealand, 86% are now urban-based. This means that for most children their primary and possibly their only experiences of nature will be urban ones.

2 Cities take up more land

As cities grow, they consume more land. Seto et al. (2011) calculate that by 2030, urban land cover will increase between 430,000 km^2 and 12,568,000 km^2 with an estimate of 1,527,000 km^2 more likely. Their study suggests that growth is taking place across large swathes of the globe; "India, China and Africa have experienced the highest rates of urban land expansion and the largest change in total urban extent has occurred in North America" (Seto et al., 2011, p. 1). Further, although the growth is unevenly distributed across the globe, the general trend, despite various urban policies favouring compact city forms, is that "urban land expansion rates are higher than or equal to urban population growth rates suggesting that urban growth is becoming more expansive than compact" (Seto et al., 2011, p. 1). The amount of non-built-on land available for wildlife and nature connection will continue to decrease.

3 Nature under stress

As urban populations grow and built land expands, nature's ability to survive and self-regulate becomes increasingly compromised. Many species and habitats will be unable to adapt to a more urbanised environment. Habitat loss or change is correlated with a pattern of species loss in cities (Shochat et al., 2010). Urban environments are normally characterised by lower biodiversity, though not necessarily by lower species numbers, as some introduced species can be advantaged by the urban setting (e.g., the grey squirrel in the UK or the Asian house gecko in Australia). Species that are less tolerant of human disturbance or those with particular habitat requirements are most impacted by urbanisation. As nature and the urban living environment becomes stressed, so too do children's lives. The health risks from environmental hazards in urban environments are considerable. Some 3 million children under the age of 5 die due to environment-related diseases (World Health Organisation (WHO), 2017). Increasingly, not only are poor environments and particularly denatured environments seen as physically harmful, but they are also problematic for children's mental well-being (Faber Taylor and Kuo, 2006).

4 Diminishing knowledge and interaction

As nature changes and biodiversity reduces, a process that has been called the 'extinction of experience' or more recently, 'environmental generational amnesia' occurs (Pyle, 1978; Miller, 2005). This is where the baseline against which an environment is assessed as an adult is founded on childhood experience of that experience, a process that can lead to a downward cycle of expectation over generations if that environment is degrading. Lack of experience also occurs in other ways. While nature is under stress in urban environments so too are children. The pace of urban life, the rise in traffic, changing social understandings of appropriate child behaviour, especially around independent movement, all act to reduce children's ability to freely navigate their urban environment and seek out opportunities to interact with nature (Freeman and Tranter, 2011). Children are finding it more difficult to experience the nature that is present regardless of its state of health.

5 Greener cities

It may seem ironic that this last point is greener cities when we have just been talking about nature under stress. Indeed, both scenarios are present often in the same city where a greening initiative in one part of a city is matched by loss of habitat elsewhere. Further complicating is the fact that green and vegetated spaces are unevenly distributed in the city, from city centres that can be severely lacking in vegetation to neighbourhoods that are well endowed in terms of green spaces and vegetation. However, even in more denuded urban landscapes, enhancement is possible. As Turner et al. (2004) demonstrate, despite this unevenness, cities overall are biodiverse, and the challenge is to better integrate biodiversity into city life and spatially across the city to enhance people's nature access. There is evidence to suggest that the appeals of Turner et al. and others to enhance and support nature are not going unheeded. Awareness of urban ecology and the natural urban environment is growing as is commitment to greening initiatives and to initiatives designed to facilitate children's connections with natural space.

The challenges discussed here are profound but not, we would argue, hopelessly insurmountable. A look at the changes in the ways that nature has and is being approached and particularly changes over time suggests that developments can and are being made to better consider children's natural relationships. Creative approaches that increase understanding of how these challenges can best be met are becoming more evident. Understanding and integrating viewpoints from different disciplinary, cultural, societal and geographically diverse settings is imperative if these challenges are to be addressed effectively.

Defining and modelling nature relationships

The relationship of people to nature is complex and challenging. The myriad forms this relationship takes are highly contested; the relationship changes over time and space and is culturally determined, and there can be a sizeable gap between stated and practiced beliefs around nature. Children's experiences of and views of nature occur within, and are shaped by, the way that nature is viewed and interacted with in the societies in which they live. We now identify some of the more prevalent viewpoints and ways of looking at the child-nature relationship. The primary relationships and means by which children connect to nature are shown through seven different lenses that reflect both traditional and more recent ways of viewing and approaching the child-nature relationship. These lenses are one suggested way of approaching understanding of nature relationships, but they are only one way and the lenses themselves are not necessarily mutually exclusive.

Dominion: In his historical assessment of "man and the natural world" in England 1500–1800, Thomas (1983) observed that belief in human ascendancy over nature was then paramount in most western societies and much of the Judaeo-Christian world and remained so until recently. This dominance, although noted for England and common to many western cultures, was not universal. From the 17th century, there were increasing reports of other cultures, such as the Jains, Hindus and Buddhists, who respected the right of creatures, even insects, to exist in their own right unexploited by people. The emergence of knowledge about these alternative positions, Thomas asserts, caused confusion amongst Britons, as such views were quite at odds with their own worldviews, which saw nature as being in need of taming and as present for human exploitation. Nevertheless, even in Judaeo-Christian religions, there were exhortations to act responsibly and with due regard to being good custodians of nature. These Judaeo-Christian views of nature as present for people to use, control and to have dominion over in many ways resonates with those present in contemporary neoliberal ideologies. Neoliberals (Castree, 2010) similarly

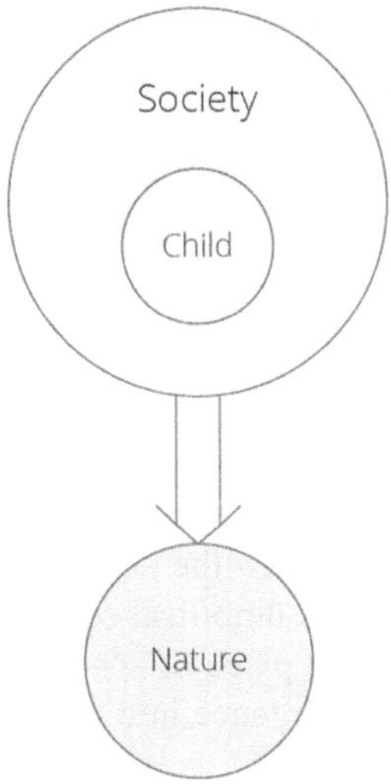

Figure 1.1 Dominion

see nature as available for human use and having economic value, though this view is increasingly set against the need to ensure that nature and natural resources are used wisely. This wise-use thinking is indicated by the increasing application of tools such as protective covenants, environmental legislation and environmental take limits designed to temper excessive use and exploitation. However, the concept of nature as present for human use remains fundamentally intact (Figure 1.1).

Integrated: In contrast, for indigenous peoples, the belief in human ascendancy is largely absent and the concept of separation between people and nature is anathema. For Māori in New Zealand, as in other indigenous cultures, the natural world is one where people are tied into the natural world, the elements of which are imbued with spiritual significance that is directly determinant of people's well-being. In Māori tradition, all things are interconnected. The Māori *whakapapa* (genealogy) includes Papatūānuku, Ranginui and their children. They were the ancestors of all parts of nature's creation story, which goes as follows (Te Ara Encyclopaedia of New Zealand, 2016):

> In the Māori creation tradition (there are numerous versions), Papatūānuku (earth mother) along with Ranginui (the sky) was born in the darkness known as Te Pō. Papatūānuku and Ranginui had several children while remaining in an embrace. The children grew frustrated with living in darkness between their parents and conspired to separate them by thrusting Ranginui above and Papatūānuku below. Thus, the world of light, Te Ao Mārama, came into being. Papatūānuku and Ranginui's many children gave birth to more children, including birds, fish, winds and water. They became the progenitors of

> all life and all natural phenomena. In this tradition, all things of this world are descendants of the children of Ranginui and Papatūānuku and all life is interconnected.

These connection beliefs are present for many Māori still today. Māori refer to themselves as "Tangata whenua – literally, people of the land – those who have authority in a particular place" (Te Ara Encyclopaedia of New Zealand, 2016).[2] This connection to the land, which begins as new-borns, occurs as follows:

> Traditionally, the whenua (placenta) and pito (umbilical cord) of new-born babies are buried in a significant place. The placenta is placed in a specially prepared receptacle and buried in a particular location. This practice reinforces the relationship between the new-born child and the land of their birth. The place where one's umbilical cord was severed is called 'te wāhi i kotia ai te pito'. This is a place of special importance for each person. It is their place of first emergence into the world, of first maturation and foundation.

This connection with nature is vital for children; they are bound to the earth physically through the placenta but also through their whakapapa (genealogy), which can be traced back to Papatūānuku and Ranginui. Although they may leave the place of their birth, they will always be bound to it through the placenta. For Māori children, as with children of other indigenous peoples whose link to the land has been disrupted following colonisation, not least through a process of urbanisation, living in accord with traditional ideals has become a greater challenge. Nonetheless, for Māori generally, these beliefs are still strong and provide a valuable and respectful way of living within the world (Figure 1.2).

Arcadian/romantic: For children, the romantic notion of natural childhoods as the norm or standard against which childhood should be measured is pervasive in much of western literature and in the various forms of the 'back-to-nature' movement. The romantic idyll sees children as innocents at one with nature and thriving in its 'beauties'. In stark contrast though, children have often also been equated with unreformed nature, perceived as 'natural beings', unformed and still driven

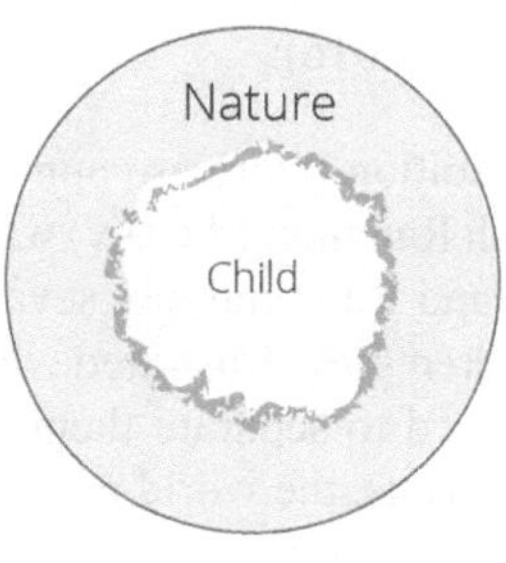

Figure 1.2 Integrated

by natural/animal instincts, whom society has yet to refine into more socially acceptable human beings. An early proponent of the romantic idyll was the poet Blake, who, in 1794, wrote *Songs of Innocence*, extolling the joys and virtues of nature. In the following poems, he contrasts the joys of rural nature with those of urban woes.

Songs of Innocence: "Nurse's Song"	*Songs of Experience*: "London"
When voices of children are heard on the green, And laughing is heard on the hill, My heart is at rest within my breast, And everything else is still.	I wander through each chartered street, Near where the chartered Thames does flow, A mark in every face I meet, Marks of weakness, marks of woe, In every cry of every man, In every infant's cry of fear, In every voice, in every ban, The mind-forged manacles I hear:

At the time Blake wrote, this romantic view of nature provided a persuasive counterpoint to the increasingly fraught and unhealthy living conditions prevailing in the burgeoning cities. William Booth, founder of the Salvation Army, echoed the thoughts of many when towards the end of the 19th century, he compared the lot of the town and country child:

> The town bred child is at a thousand disadvantages compared to his cousin in the country . . . To rear healthy children you first want a home; secondly milk; thirdly, fresh air; and fourthly, exercise under green trees and blue skies.
> (in Cunningham, 2006, p. 185)

The concern for many social reformers like Booth was the poor living environment of urban children. Certainly, in the move to industrialise, little thought was given to consciously including nature in the urban fabric (Figure 1.3).

Denatured: More recently, the concerns for urban children in most western countries have shifted from the issues of survival and serious poverty identified by early social reformers (although these still exist) to those more related to emotional

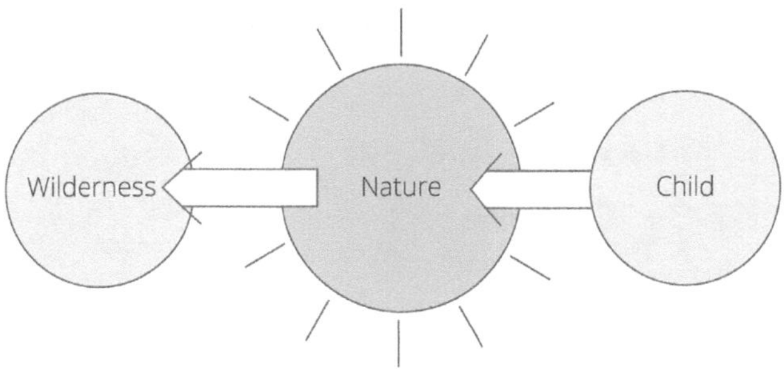

Figure 1.3 Arcadian/romantic

and spiritual well-being associated with loss of nature connectedness. One of the questions we address in this book is the following: What biodiversity and potential for nature-based experiences are available in urban areas? Concerns around reduced nature connection are ongoing but perhaps for different reasons from those historically. One of the foremost proponents of the nature connection movement is E. O. Wilson, who identifies a clear relationship between nature connection and well-being described as "the innately emotional affiliation of human beings to other living organisms" (1993, p. 31). Childhood is seen as pivotal in realising biophilia:

> If by some fairly young age, however, nature has not been experienced as a friendly place of adventure and excitement biophilia will not take hold as it might have.
>
> (Orr, 1993, p. 428)

The growing separation between children and nature is seen as being especially prevalent amongst urban children. Concern about this separation has fuelled a growing movement of 'back-to-nature' advocates (Figure 1.4), the best known proponent of which is Richard Louv (2005) and the Children and Nature network (Charles and Louv, 2009). Louv coined the term 'nature deficit disorder' (NDD), where lack of contact with nature is seen to be associated with health disorders, such as attention deficit hyperactivity disorder (ADHD) / hyperactivity, obesity, stress and depression. A more recent movement is the similar UK Wild Network, which explains itself as follows:

> We are the wild network. A growing movement of people on a mission to re-wild childhood. Together we want to help kids roam free, play wild and lead nature-rich lives. JOIN US. For the urban environment, the prognosis is seen as troubling. Those edgelands, the bits of wildness not far from every home are vanishing fast.
>
> (The Wild Network)

These child-nature advocates focus on the damage being done to children through lack of nature connection and promote a 'back-to-nature' movement to reconnect children with the outdoors. Some movements have become more formalised as in

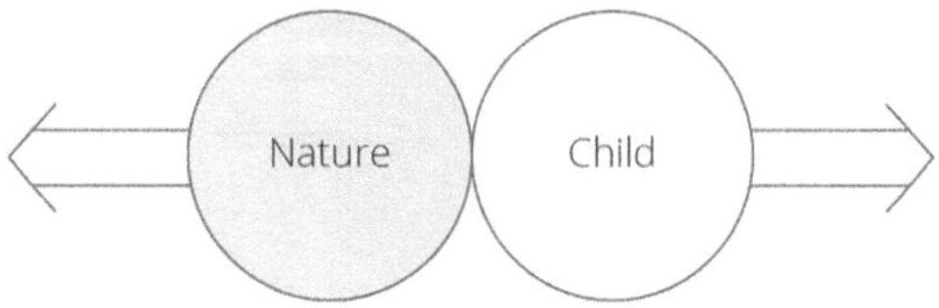

Figure 1.4 Denatured

the US No Child Left Inside Act of 2008, passed by the House of Representatives on September 18 but never voted on in the Senate (Congress). The back-to-nature movement, with its focus on urban biodiversity deprivation, has been very active, and its message, especially around addressing 'nature deficit disorder', has successfully caught the public and academic imagination. However, while the principles of the movements are well intended and important, their implementation adopts an approach that has vestiges of the 'wilderness' attitude to nature, directed towards seeking out the pristine, the rare and the rural, albeit often within an urban-type context. This lens is strongly reminiscent of the rural idyll as portrayed by the early romantic poets, such as Wordsworth.

Professional/directed: In understanding nature connection and the reasons for its presence or absence, professionals such as planners and scientists can be important in defining and framing the form of nature in the urban environment and the ways children are able to experience nature. The professionals act as conduits through which nature and children interact. Through their actions, they can advance particular ways of presenting, interpreting and moulding both nature and how children relate to nature. Environmental educators turn their professional gaze to teaching and facilitating children's access to biodiverse areas usually through a raft of formal in-nature learning experiences. Formalised education experiences though can only ever be a small and special part of children's lives and experience. Teachers and other adults in children's lives who support a strong environmental programme can enhance children's ability to connect to nature. Further, as children's freedom and free time is curtailed, access to nature is becoming increasingly directed as in an environmental education experience, a school trip to a nature reserve, a family outing or a holiday camp. The nature interaction becomes directed by an adult intermediary rather than being a self-directed exploratory and interactive child-led experience (Figure 1.5).

Filtered: For many children and for many cities, the issue is not so much a lack of nature as an inability to access the nature that is present. Access to nature is mediated through and controlled by a number of often powerful gatekeepers.

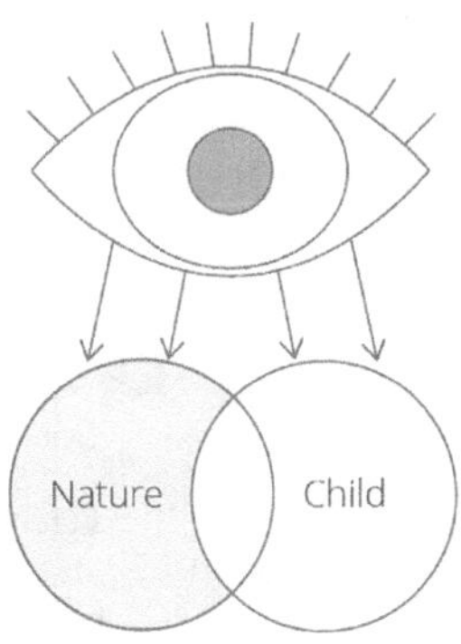

Figure 1.5 Professional/directed

These include social gatekeepers, in so far as certain children do not feel welcome in certain spaces, perhaps because the residents are not perceived as friendly. The children in our *Natural Neighbourhoods for City Children* study living in a low-income state-housing area rarely accessed the local beaches that were located in the adjacent higher-income suburb despite these beaches being within walking distance of their homes. Similarly, access can be filtered through constraints, such as being unable to afford a home with a garden (economic), living in an area deficient in biodiversity (location), living in an area which is deemed unsafe (social) or living where girls are expected to stay within the family home rather than be playing outside (cultural) or where parental fears are generated by news stories reporting attacks on children playing outside or castigating parents for being neglectful following a child's death from drowning in a river (media). Understanding these contextual factors that mediate nature access is fundamental to explaining children's degree of nature connection (Figure 1.6).

Everyday nature: What we concentrate on in this book are those nature connections and experiences that can and should be part of everyday life (Figure 1.7) by looking at the nature that is present and can be supported within children's urban lives. Some nature experiences will be special, such as a visit to a nature reserve or ecosanctuary, and often are located outside the children's neighbourhood. Others will be the everyday: planting vegetables; removing spiders, flies, ants and other sundry creatures from the house; playing on the grass; climbing trees; messing in the local stream; walking the dog or walking to school in inclement weather, such as snow or rain. These existing and potential experiences of nature within local

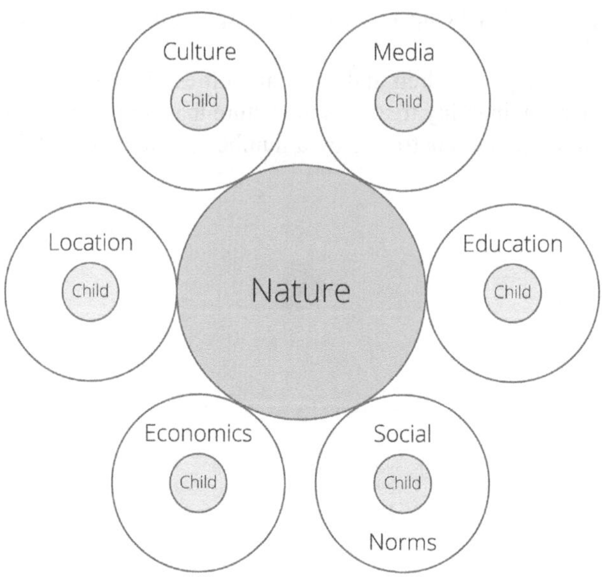

Figure 1.6 Filtered

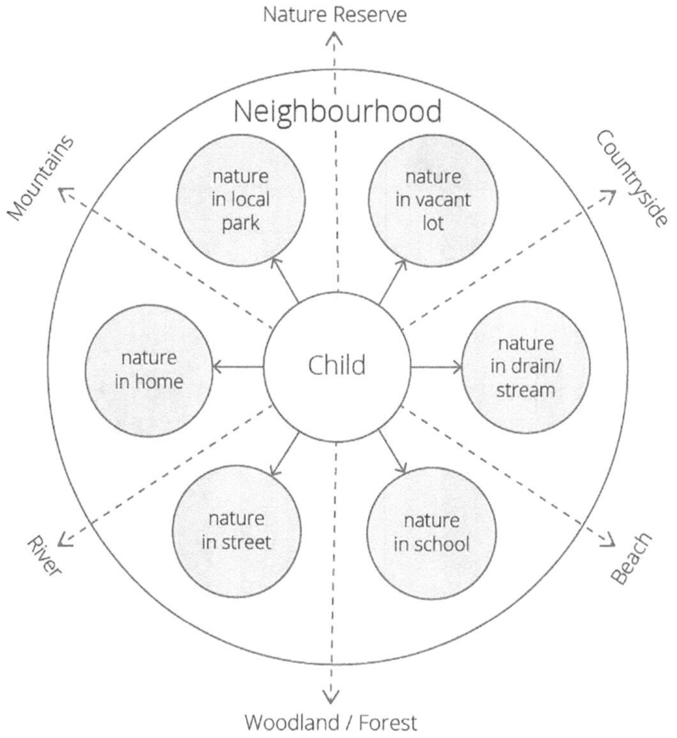

Figure 1.7 Everyday nature (*the approach used in this book*).

neighbourhoods are becoming the foci that are more common for urban ecologists. Their work is increasingly challenging the denatured view of cities (Goode, 2014; Marzluff et al., 2008) and celebrates the often admirable ways that wildlife adapts to this new environment. However, attendant with this admiration comes also the need to protect and support the wildlife that fails to survive and for which a different approach is needed.

How nature is viewed is important, as it determines the ways that nature connection is evaluated and responded to by researchers, professionals, parents and children themselves. If children are led to believe the wildlife they experience is of little value, perhaps because it is a common introduced mallard rather than a more important native 'paradise duck', then this will influence how they value the species and its habitat. Similarly, a strong neoliberal approach, which values the local forest only with respect to the value of its saleable timber and attendant jobs, can influence children's own assessments and ways of connecting to that environment. Conversely, adopting an indigenous view would mean the child sees his or her own well-being as directly connected with that of the forest. In this book, we particularly address the 'denatured' lens by critically assessing the veracity of this

The urban green elements continuum					Non-urban wild
Indoor pot plants - totally bulit environment, minimal greenspace	Built dominates, small greenspaces, grass, few trees, ecologically poor	Mixed built-green, some garden space, trees, some greenspaces with ecological interest	Green dominant but still a built environment, range of greenspaces of varying ecological quality	Green dominant, range of greenspaces most are ecologically rich, predominantly natural ecosystems	Wilderness - no built element

The urban nature access continuum					Non-urban
No greenspace access	Minimal access to limited greenspace, e.g. lawns	Some accessibility to formal/domesticated nature places e.g. parks and ducks	Free access to wilder but still managed nature e.g. natural tree stand in park, pond	Ecologically rich habitats, primarily natural e.g. woodlands, wetlands	Wilderness

Figure 1.8 The nature continuum.

currently very influential approach to viewing children currently growing up in cities today.

Children's nature experiences and interactions take place along a continuum – the nature continuum (Figure 1.8) that moves from very limited access and experience, perhaps in the form of pot plants at home or grass and street trees in the neighbourhood, to ecologically deep and meaningful experiences in natural habitats. Those children living in very dense inner cities, in compact informal settlements, in high-rise or other naturally constrained environments may find their natural experiences limited. Even so, their limited experiences are still valuable (e.g., sitting under a street tree) and should be respected rather than dismissed. The challenge is to enable children to move along the continuum, to develop meaningful interactions with the nature that is available no matter how limited and to provide more natural experiences at the greener end of the spectrum.

Much of the literature on children and their natural environments is negative in the sense of proclaiming a generation who are divorced from nature and for whom this divorce is creating major emotional and physical trauma. The focus has been on experiences at the green end of the continuum only, disregarding other experiences along the continuum. Whilst conceding that children do need nature, we offer a more positive view and identify how children in cities can and do connect with nature and how cities can and do provide meaningful natural experiences. We argue that children are resilient and given the right social and physical environments will intuitively make natural connections. Relationships with nature are becoming seemingly more difficult for children to achieve, but we argue instead that these difficulties can best be addressed through adopting a more positive response, one that enables them to connect emotionally and physically to biodiversity and natural places.

Positive orientation

The book focuses on cities and the opportunities they supply for nature interaction. For those cities, or spaces in cities, where such opportunities are lacking, it offers suggestions on how such deficits can be remedied. However, there are few cities that offer no opportunities for nature interaction; even the most densely built cities have some street trees, waterways, roadsides, parks (albeit of variable quality), pot plants and other domestic greenery. Even in densely built-up cities, such as central Seoul, nature is present (Figures 1.9–1.14) and surprises happen, as when two girls found a rabbit in a small green patch by the old city wall (Figure 1.9). The rabbit was clearly not wild, and how it got there will remain a mystery, but the memory of that interaction will likely remain with the children for a long time. The other figures show a tree surrounded by buildings in downtown Seoul, pot plants in a traditional housing quarter, a school trip and a family trip to a park, and examining the ground during a visit to a palace. These figures show the affinity that people have with nature, seeking out opportunities to bring nature into even inhospitable environments and engaging with the nature that is available in nearby parks and historic places.

Figure 1.9 Finding a rabbit in bushes by the city wall in central Seoul.

Figure 1.10 A cherry blossom tree 'hidden' in the middle of buildings in central Seoul.

Figure 1.11 Bringing green into the home environment, pot plants outside a traditional home.

Figure 1.12 School outing to the Experiential Learning Park at Seoul Forest.

Figure 1.13 Family time in the park.

Figure 1.14 Checking out the earth at Deoksugung Palace in Downtown Seoul.

The focus in the book then, is on the myriad ways that nature makes its presence in the city, particularly how it enters into ordinary everyday lives. These nature interactions are examined through a combined theoretical and experiential lens. Focusing on nature and especially 'nature interaction' is a challenge, not least because notions of nature are variable and can be highly context-specific. This variability gives rise to often vigorous debates around questions such as what constitutes 'authenticity' and what we should 'value' and debates around 'good' and 'bad' or appropriate and inappropriate behaviour in nature. These questions permeate strongly discussions around types of children-nature experiences with some experiences encouraged as formative, while others, often the everyday experiences, sometimes dismissed as valueless or synthetic.

The case for natural urban childhoods

We advocate the case for natural childhoods within a context of real lived experiences and with reference to current urban environments internationally. This book combines the work of and the benefits that come from the fields of expertise of the two authors – namely children's geographies / planning and ecology. It adopts a multidisciplinary perspective that embraces the natural and built environments and the challenges for both in supporting nature and children's connections to nature. The content combines knowledge and lessons from the natural sciences about the city and its habitats, species and environments together with the role of geography and planning in understanding children's spatial lives in order to assess and create better and more natural environments.

The nature-child relationship is explored through four parts. Part I sets the context, exploring concepts around the child-nature relationship and the different ways this relationship can be approached. In any exploration of children's worlds, the voices of children need to be central, as they have the purest insight into their own ways of seeing and experiencing. In the second chapter, children's voices are reflected, and, in doing so, this enables children's voices to be heard and considered at the outset, setting the context for the discussions that follow in the remainder of the book. Part II examines the physical context of the city, its wildlife and the species it contains in relation to people. In chapters 3 and 4, the city is assessed from an ecological perspective, detailing the ways biodiversity occurs in and adapts to the urban environment and its human impacts and relationships. Chapter 5 explores the ways cities are planned, change and develop and the ways the green movement is changing cities for the better. Part III considers the different contexts and influences on nature interactions. Chapter 6 examines families, nature and the home; chapter 7 nature and health; chapter 8 culture and difference in relation to nature and chapter 9 mediated nature, which identifies the way different forms of media, story-telling books, films, and computers, all play a role. Part IV looks to the ways that better, more natural cities can be created and children's access to nature supported. The focus initially in chapter 10 is on challenging environments, primarily high-density and deprived environments. Chapter 11 goes on to explore the roles of governments, professions and those who play a role in creating biodiverse spaces for children at national and local levels. It provides examples of how spaces that form key spaces in children's lives, such as playgrounds, schools, hospitals and parks, can be designed or redesigned to enable nature interactions. While the emphasis is on professionals, it also identifies how children can be included in providing the data necessary to improve our understanding of city biodiversity. Chapter 12, the final chapter, concludes by reiterating the need to create biodiverse spaces in children's lives and to provide children with the resources and freedom they need to access these spaces.

This book differs from more 'standard' children and the environment texts, as although we recognise and investigate this child-environment relationship, we also analyse the available nature from a scientific perspective and with a view, through planning, to identify positive practical processes to support child-nature relationships in cities. We build on our extensive previous research to enhance understanding of children's environmental relationships. The book identifies practical ways that nature can be enhanced in children's lives and in the cities they live in, through both scientific endeavour and better planning and urban design processes. The third contribution after the two authors comes from children themselves. Children contributed to this book through their schools. We worked with children in two schools where they discussed and explained their own nature interactions. We also include the views from children in our *Natural Neighbourhoods for City Children* study. Their words and views form an integral part of the debate and provide valuable insights into children's own nature connections in the cities in which they live.

Notes

1 The mallard has caused the virtual extinction of the native grey duck in New Zealand through hybridisation.

2 Te Ahukaramū Charles Royal. 'Papatūānuku – the land - Whenua – the placenta', Te Ara - the Encyclopedia of New Zealand, http://www.TeAra.govt.nz/en/papatuanuku-the-land/page-4 (accessed 9 March 2018).

References

Castree, N. 2010. Neoliberalism and the biophysical environment: A synthesis and valuation of the research. *Environment and Society* 1(1): 5–45.

Charles, C., and Louv, R. 2009. *Children's Nature Deficit: What We Know: And What We Don't Know*. www.childrenandnature.org/.

Congress.gov. 2016. www.congress.gov/bill/110th-congress/house-bill/3036.

Cunningham, H. 2006. *The Invention of Childhood*, BBC Books, London.

Faber Taylor, A., and Kuo, F.E. 2006. Is contact with nature important for healthy child development? State of the evidence. In C. Spencer and M. Blades (Eds.), *Children and Their Environments*, Cambridge University Press, Cambridge, UK, pp. 124–140.

Freeman, C., and Tranter, P. 2011. *Children and Their Urban Environment: Changing Worlds*, Earthscan, London.

Freeman, C., van Heezik, Y., Hand, K., and Stein, A. 2015. Making cities more child-and nature-friendly: A child-focused study of nature connectedness in New Zealand cities. *Children, Youth and Environments* 25(2): 176–207.

Goode, D. 2014. *Nature in Towns and Cities*, Harper Collins, London.

Louv, R. 2005. *Last Child in the Woods: Saving Our Children from Nature-Deficit Disorder*, Algonquin Books, Chapel Hill, NC.

Marzluff, J., Shulenberger, E., Endlicher, W., Alberti, M., Bradley, G., Ryan, C., ZumBrunnen, C., and Simon, U. (Eds.). 2008. *Urban Ecology: An International Perspective on the Interaction between Humans and Nature*, Springer, New York.

Miller, J.R. 2005. Biodiversity conservation and the extinction of experience. *Trends in Ecology and Evolution* 20(8): 430–434.

Nature Conservancy UK. www.nature.org/ourinitiatives/urgentissues/ accessed 19–2–2016.

Orr, D. 1993. Love it or lose it: The coming biophilia revolution. In S.R. Kellert and E. O. Wilson (Eds.), *The Biophilia Hypothesis*, Island Press, Washington, DC, pp. 415–440.

Pyle, R.M. 1978. The extinction of experience. *Horticulture* 56: 64–67.

Seto, K.C., Fragkias, M., Guneral, B., and Reilly, M.K. 2011. A meta-analysis of global urban land expansion. *PLoS ONE* 6(8): e23777. DOI: 10.1371/journal.pone.0023777

Shochat, E., Lerman, S.B., Anderies, J.M., Warren, P.S., Faeth, S.H., and Nilon, C.H. 2010. Invasion, competition, and biodiversity loss in urban ecosystems. *BioScience* 60: 3.

Te Ara Encyclopaedia of New Zealand. www.teara.govt.nzaccessed 2–2–2016.

Thomas, K. 1983. *Man and the Natural World*, Changing Attitudes in England 1500–1800, Penguin, London.

Turner, W.R., Nakamura, T., and Dinetti, M. 2004. Global urbanization and the separation of humans from nature. *BioScience* 54(6): 585–590.

The Wild Network. www.thewildnetwork.com/vanishing_green_space accessed 19–2–2016.

Wild Network. www.thewildnetwork.com/thewildnetwork accessed February 2016.

Wilson, E.O. 1993. Biophilia and the conservation ethic. In S.R. Kellert and E.O. Wilson (Eds.), *The Biophilia Hypothesis*, Island Press, Washington, pp. 31–41.

World Health Organisation. 2017. *Children's Environmental Health: The Environment and Health for Children and Their Mothers*. www.who.int/ceh/publications/factsheets/fs284/en/

WWF. 2012. *Urban Solutions for a Living Planet*, WWF International, Switzerland.

2 Children's views and perspectives

Listening to children: their views on the nature relationship

Children have been the focus of much concern regarding their connection and disconnection from nature. There has been much written on the benefits of children's connection to nature and on the problems that occur should this connection be absent. However, in the articulation of the benefits and problems associated with the child-nature relationship, the voices of children are conspicuously absent. If children's voices are not heard, then how can we as professionals, whether it be as teachers, ecologists, planners, conservationists, decision-makers or environmental managers understand and provide for children's needs in connecting with nature? Further, children today inhabit the world that the adults have shaped and continue to shape. If this world is to be shaped in ways that enrich children's lives, then adults have to reconsider and take responsibility for a world that seems to be increasingly frustrating children's ability to inherit a nature-rich environment.

In this chapter, we profile children's views, perspectives and experiences of nature. We use children's voices as expressed in our own research projects and present the views of children interviewed for this book. The main research project used in this chapter is the *Natural Neighbourhoods for City Children* study. This study involved 187 children attending nine schools in three New Zealand cities. The children were aged 10–11 and lived in neighbourhoods that were characteristically low, medium and high socio-economic areas. In the larger cities, these neighbourhoods are ethnically quite diverse. In addition we interviewed children from two schools in the lower South Island of Dunedin, a city of some 120,000 people. The two schools are Dunedin North Intermediate (a class of 12–13 year olds) and North East Valley Primary School in Dunedin (a class of 6-year-olds). Children attending these two schools span the socio-economic range from low to high, and although the schools are urban, some children live on the urban/rural interface. We fully recognise that these three sources are a small sample and do not intend the voices of the children in this book to be representative of children generally or even of children from the neighbourhoods our children live in. However, the children in this book do present their views and provide important insights into the everyday lives of children and their natural encounters, and they also provide some indication of what children think about both nearby and more distant nature.

Children have a right to be heard. In this respect, The United Nations Convention on the Rights of the Child (UNCRoC) (UN General Assembly) is important. It includes articles relevant to children's voices and participation. Article 3 articulates the need to consider 'the best interests of the child'. In this book, we put forth the case that access to nature is a fundamental right for children and one whose absence is injurious to their well-being. Further, Article 31 recognises children's 'right to play', and again we would argue that play, in the form of play in natural spaces and the ability to be able to explore and have the freedom to access natural places, is an essential right. This chapter pays particular attention to Article 12, which recognises the child's right to be heard on matters that affect them, and the natural environment is patently one very important such matter. The articles in fuller detail are as follows:

> **Article 3:1** In all actions concerning children, whether undertaken by public or private social welfare institutions, courts of law, administrative authorities or legislative bodies, the best interests of the child shall be a primary consideration.
>
> **Article 12:1** States' Parties shall assure to the child who is capable of forming his or her own views the right to express those views freely in all matters affecting the child, the views of the child being given due weight in accordance with the age and maturity of the child.
>
> **Article 31:1** States' Parties recognize the right of the child to rest and leisure, to engage in play and recreational activities appropriate to the age of the child and to participate freely in cultural life and the arts.

Although there are few who would argue against UNCRoC or children's rights, in general, there remains limited understanding of how children should be listened to and how children's rights under the convention can be operationalised. How to listen to children is seen as particularly difficult when the other competing (adult) voices and views can be much more easily accessed and far less easily ignored.

In relation to the growing concerns about possible nature disconnect, a few prevalent societal concerns dominate the discourse: concerns around lack of knowledge about nature; children's perceived preference for sedentary, usually electronic play over outdoor play; children's declining physical activity; their more time-pressured lives and safety issues. Children through this discourse are positioned as unhealthy recipients of a modernising, alienating and less-natural world. Yet, it is seldom asked whether this discourse reflects how children see their world and their natural relationships within it. Children need a platform to be able to input their views directly and, thus, themselves influence the discourse around child-nature relationships. In 2010, the New South Wales (NSW) Government produced an excellent guide to child-friendly community building, *Built for Kids*, which is useful for involving children in environmental development. The book gives the following reasons for listening to children:

- To ground adult understanding and decision-making in the reality of children's experience
- To remove the need for adults to assume who children are and what they want
- To recognise children's right to participate in community development processes
- To recognise children's contribution
- To recognise that the knowledge and experience of children differs from that of adults

(after NSW Commission for Children and Young People, 2009)

In this chapter, we ask, how do children see their natural world? We profile research that has directly engaged with children's voices and children's own direct articulations and input. We begin by briefly identifying alternative methods for enabling children's voices to emerge and then address a few key questions:

What is nature and how do they feel about nature?
What is the relationship between nature and technology?
What do they see as the challenges?
What are the questions they think should be asked?
Where and how do children encounter nature?

Some excellent progress has been made in enabling children to have a say. However, in general, children's voices and participation remain marginal, and for some groups, such as very young children, children from some ethnic groups, recent immigrants and disabled children, exclusion persists as the norm. Participation and the right to be heard should not be dependent on 'possession of a set of identifiable skills or attributes' (Hill et al., 2004, p. 85), or being in an accessible location. Under UNCRoC, it is an inalienable right for children in the 196 signatory countries (the US is not a signatory to date) to be able to express their views on matters that affect them.

Accessing children's views and perspectives: some methods

Techniques can and have been developed, especially post-UNCRoC, with a view to better understanding children's lives and for including children in participatory decision-making. Some of the more commonly used methods include the following:

1 Verbal: interviews, focus groups, stories and storytelling
2 Representations: drawings, maps and models

3 Electronic: videos, simulation games, geographical information systems (GIS) and global positioning systems (GPS)
4 Site-based: Photo voice (photos of specific sites), site visits and go-along walks

With any technique, there are benefits and limitations, and the technique is at best just a vehicle for enabling partial access to children's knowledge and views. A study of children's neighbourhoods found maps drawn by children varied in what they showed of children's local knowledge, depending on children's interest and map-drawing skills. It was therefore concluded that it is unwise to infer that maps represent all of children's knowledge (Freeman and Vass, 2010). Drawings are nonetheless still valuable, especially when used with pre-literate children or children who struggle with literacy. They are a good way of initiating conversations with children with the purpose of enabling them to show their knowledge of and views on their neighbourhood environments (Figure 2.1). The following drawings (Table 2.1) by children aged 6 from North East Valley School provided information on what children knew and thought about nature and acted as an initiator for discussions on nature.

Figure 2.1 A researcher in Fiji uses a child's drawn neighbourhood map as an aid in discussing her experiences of living in Suva.

Table 2.1 Using drawings to elicit young children's views on nature

Drawing	*Comment and discussion point*
	Pets, including rabbits, and the outdoors are commonly featured in children's drawings, in this case 'bunnies in a cage'. In discussions with children pets are frequently mentioned and do provide children with direct experience of the natural world and the need to nurture animals through looking after them.
	This drawing shows a cheetah mother protecting her baby, which led to a discussion on how animals protect and care for their young.
	This boy, when asked what he liked about nature, said 'everything' and drew his family and their activities as part of nature. We have found that it is common for children to see themselves as being part of nature rather than as a separate entity. In this sense they affiliate with the 'integrated lens' way of seeing nature (see Figure 1.2).
	This girl liked nature, as there are trees to climb. She is worried about chopping down trees and included one in her drawing. Her drawing evidences early environmental concern.

Drawing	*Comment and discussion point*
	This drawing represents the various activities and nature available in the school grounds and neighbourhood. The house is clearly part of a natural environment. This class had been learning about butterflies, and the children were very keen to talk about butterflies and how they used their closed wings as camouflage.
	This drawing shows species awareness with drawings of different kind of birds in New Zealand; kiwi, fantails and others. The children were asked what is not good for nature, and this boy added a pig. In New Zealand, introduced 'wild' pigs (boar) are used for hunting. They are problematic predators on native species and damage forest habitats. Also present in the drawing are worms as a food source. The tall green stems are ferns, a prominent native plant and one that is culturally important as it is often used to represent New Zealand identity, as in the national rugby team 'All Black's' silver fern motif.
	The koru (unfurling young fern) is an important symbol of new life in New Zealand and one used in many settings. This boy uses it in its natural setting as unfurling ferns providing habitat for a kiwi and another unidentified native bird flying overhead.
	Kiwis are rare throughout much of NZ, and as this drawing indicates, are present in natural areas such as in the mountains. Many of the children drew kiwis, a species that few if any would have seen. However, kiwi are iconic New Zealand species and the children clearly recognised their importance, knew they are threatened by introduced species such as dogs and pigs, and knew they rely on insects for sustenance. Again, insects appear in relation to kiwis, showing evidence of understanding of species' relationships.

Children's voice

What is nature?

We asked this question of children from the two schools contributing to this book and of children in the *Natural Neighbourhoods for City Children* study (Table 2.2). Some children found this a difficult question.[1] The first response was often to mention living things in nature, such as birds and trees. Then especially from older children, there was an attempt to convey complexity and to define

Table 2.2 What is nature? Selected responses[2]

6 year-old children	*10-13 year-old children*
Birds: kiwi, fantail, tui, wood pigeon, house sparrows, silvereye, grey warbler, goldfinch, greenfinch, kaka-parrot.	Trees plants, all the things around us, but not electronics or roads, not man-made things.
Spiders! Tarantulas, but only in Australia!	Trees, grass, flowers, can also be animals, bees, birds and things, lots of things, pretty much almost anything outside, unless man-made.
Possums, sheep.	Most things that live, rocks, caves and natural stuff, stuff that is naturally made.
Colourful frogs, colourful frogs are poisonous.	Everything green you can find outside.
Big trees.	Nature is animals, and mother nature looks after all the nature, and plants and other things that need to be watered because that's what mother nature does. Living things and Earth.
Volcanoes on TV!	Gentle place with lots of trees and bushes and flowers, and lots of animals live there like native birds like kiwis and tuis.
The playground has heaps of nature! Trees . . . and bushes and birds and worms.	Nature is outdoors and it can be living.
My rabbits and guinea pigs!	Nature is a lovely thing.
The grass outside my house.	Things that are alive, except humans and pets are nature.
Snow on the hill.	Nature around the neighbourhood is stuff that can move and live.
Caves and whirlpools.	Nature is all living things.
Cats and possums . . . But they eat kiwis.	Plants, trees, and the environment around us.
Stoats, dogs, birds, stones-predators who eat kiwis.	Nature is something that has been here for a long time, and is usually green.
A long list of species including: snails, flowers, butterflies, cherry trees, birds-seagulls, parrots, slugs, spiders, moths, daddy long legs, frogs, snakes, ducks rivers, mountains, stones.	Can't live without it, would also say it includes most of the stuff we need to live, like food and air, and it's got lots of animals.
Blue flowers, pink flowers-honey bees like them, white flowers.	Life.
Eagles! They are everywhere! Except New Zealand.	Nature is what grows by itself, nature is things that produce oxygen and live, and you can't just destroy it.
Fish in the beach and the water in the river, like goldfish	
Colourful frogs, "colourful frogs are poisonous."	

nature through contrasts (natural-built), expanding to talking about relationships between natural elements and people and nature. Younger children seem to focus on species and are keen to name them and to show wider knowledge of nature – as in poisonous frogs, eagles and predators. They show relatively little discrimination between what is local and what is distant (e.g., from a different country) or what is native and what is introduced. The same child drew both the rabbits in the cage and the drawing of the cheetah, demonstrating a wide awareness of very different nature, moving easily in discussion between these two very different natural elements. The responses from the older children were more reflective of nature as an entity that encompasses multiple elements, with several references to all living things and life in general. One 6-year-old also talked of nature as everything including people (see the third drawing in Table 2.1). No older children included people in their definitions of nature; rather, they made a clear distinction between natural and man-made. What also became clear in the older children's responses is the articulation of emotions – nature as peaceful, gentle and vulnerable – with some children expressing responsibility for nature.

The children went on to talk about their feelings about nature, and for the 6-year-olds, this was about what they liked and disliked or what they considered good and bad in nature. Several themes emerged. Firstly, the importance of family pets: *I like my cat/rabbits/dog* (Figure 2.2). They appreciate nature's functionality in *I like climbing trees / rolling on grass / hiding in bushes*. Again, there was a tendency to list species they liked: *ducklings / koru / teeny weeny spiders / flowers / ladybugs*. Opinions on bugs and worms ranged between like and dislike. Dislike was also qualified by recognition of their value as in *Worms are good because they eat the soil* and *I like in nature that bugs stay alive*; however, spiders were universally disliked. The children were concerned about death and did not like to see dead

Figure 2.2 Children recognised species are complex as in this quote about cats: "Although my cats kill birds, they also kill other pests like rats, mice, and even a baby possum once". Cats featured in the children's quotes more than any other domesticated animal.

animals, even worms or bugs; *I don't like to see dead bugs and birds. I once saw a dead rat on my lawn.* They were concerned about kiwis being eaten by dogs. They found it far easier to talk about what they like or is good than the negatives.

In the *Natural Neighbourhoods for City Children* study, we wanted to investigate how children perceived their own neighbourhood nature in terms of availability. This question derived from the finding by Louv (2005) and others, suggesting children were insufficiently connected to nature. New Zealand cities, by international standards, are biodiverse (Clarkson et al., 2007; Freeman and Buck, 2003) in the sense of being well vegetated. We were interested to see if this was how children perceived the city. We asked two related questions: "Is there a lot of nature?" and "Is there enough nature where you live?" Most children said there was a lot of nature and that there was enough (Figure 2.3). Interestingly, children from the higher socio-economic group were more likely to say there was not a lot of nature even though biological surveys suggest they have access to more biodiversity compared to children from low socio-economic groups. Lower biodiversity encountered by children from low socio-economic groups was generally a result of lower private yard/garden biodiversity rather than public park biodiversity. Children from the higher socio-economic group tended to have more mature garden vegetation.

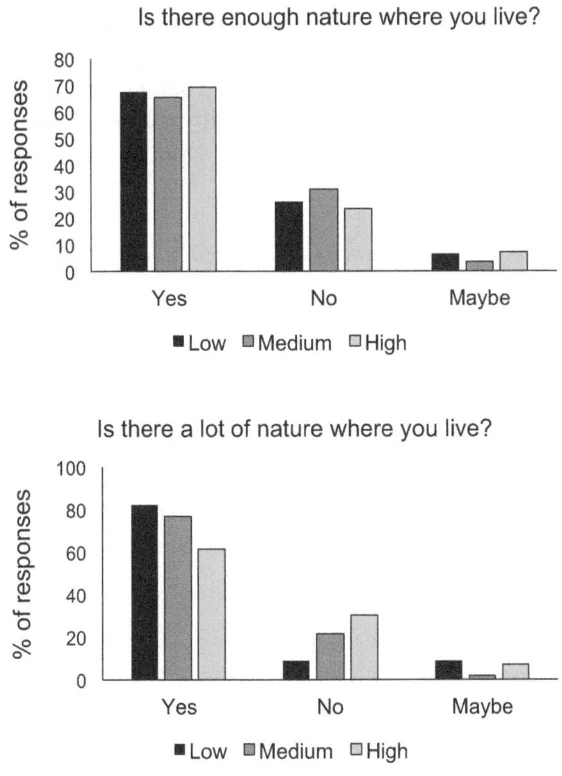

Figure 2.3 Responses from 187 children from low, medium and high socio-economic households to questions about nature and the neighbourhood.

The technology-nature relationship

Technology has been vilified in the child-nature debate as being responsible for removing children from nature. In this book, we identify a number of pressures children are experiencing that make it more difficult for them to experience nature, especially through outdoor play, and argue that technology is but one of several impacting factors. In 2003, Holloway and Valentine found the influence of technology to be less than expected and observed that most children (weather and seasons permitting) preferred to play outdoors. This led them to conclude that the technology/computer preference was not necessarily emanating from the children themselves. Since their book was first published, there has been a steady rise in the availability of technology. In the UK, 34% of children aged 5 to 15 years have their own tablet/computer, rather than using devices belonging to their parents or school, and six in ten (62%) children use a tablet at home with access continuing to rise (OfCom, Ofcom is the communications regulator in the UK). Childwise reports that children aged 5–16 now spend six and a half hours a day using Internet-connected devices and watching television (TV) (Childwise UK, 2016). We asked the 6-year-old children about technology and nature. Several talked about technology as an aid: *Yes, you type and it will come up with the animal you are looking for and it will tell you things. You can google, you can search up stuff,* and *You can search up nature and it will tell you things.* Some children also recognised the role of books: *nature books like spider books, spiders are nature, redbacks [a type of venomous spider] and bugs* and *Yes, I have a nature encyclopaedia and I look in it a lot.*

The older children were asked more specifically whether technology helps them experience nature or not. As with the adults, some children reflected concerns about technology:[3]

> *Because of technology no one is actually going outside because they can do everything sitting in front of a screen. Nowadays, for example, people don't need to actually walk one of the great walks when they can just see it all on Google Maps.*
>
> *Tech keeps you inside and you get fat and eat Doritos [potato chips]. But if you go outside you can work out and not be like a fat lazy person who is obnoxious.*
>
> *No, it stops people from experiencing nature outside of screens.*
>
> *"Definitely not!" For most people it stops them from going outside because they would rather sit on the couch on their phones all day.*

Other children saw the positives in technology:

> *Yes, because you can research stuff.*
>
> *Yes, because of simulations and sometimes Google Maps. It helps you learn more about it like we learn more about space in the Planetarium.*
>
> *Documentaries and stuff help people learn about faraway stuff.*
>
> *Yes, technology is everything it helps you find out about nature and takes you to nature as well.*
>
> *Yes, because technology helps us analyse nature.*

However, for some children, there is no definitive answer, as they recognised the need to balance the positives and negatives from technology:

> *It's sort of a yes and no question. Yes, because it gives you ideas of plants and things. No, because video games keep us inside.*
>
> *Most of the time no, but technology like cars help you to get to nature, and photographs can also help. When sitting indoors, technology won't help, though.*
>
> *Most of the time NO. But transport can help get you to nature. However, computers Do NOT.*
>
> *What technology? Games like Pokémon do get people out, but they don't pay much attention to their surrounds. Cars help people get to nature. Most computers make people stay inside.*
>
> *Yes and no. Technology keeps people inside but also educates people about nature and can motivate people to get outside.*

Technology is an integral component in children's lives. These children have grown up in a technology-available world, accessing it at home, at school, and in the wider world. Perhaps the response from one child sums up the argument from a child's perspective: *Tech is nothing. Deal with it.* The question for the adults is how to do that so that nature is not a loser in the battle but an aid.

What are the challenges? Children's own concerns

For the children in our *Natural Neighbourhoods for City Children* study and the 12–13-year-olds, there was considerable concern expressed about the need to care for nature; they also recognised threats to nature and expressed concerns at two scales, the global and the local.

> Global concerns:
> *Pollution. Methane. Cars. Animals. Humans. All that stuff.*
> *Humans because of technology development and an expanding population.*
> *Yes, pollution destroys our environment and animals and smells.*
> *Yes – there is pollution and pests eating natives.*
> *Pollution – destroying plants.*
> *Yes because what we do affects the environment in a bad way which is causing global warming.*
> *Yes, there is . . .*
>
> > *Cattle farming produces methane*[4] *also pollutes waterways.*
> > *Pollution – increases global warming.*
> > *People over-fishing – gets rid of species.*
>
> *Yes. Global warming can destroy nature. And possums attack bird nests.*
> *Yes, nature's main enemies are humans and nature itself as humans cause pollution and destroy nature for their own benefit, meanwhile animals are slowly destroying plants.*

I don't think there are many threats but there are some major things such as pollution and global warming caused by factories producing smoke and cows and other things like that.

Cows – producing methane – global warming.

What we eat, what we live in, and what we wear is all used or affected by nature.

Some poor countries destroy nature for money e.g. Cutting down rain forests.

In Aboriginal Australia they draw and celebrate nature.

Every country has their own way with dealing with nature. In some countries you can't see the sky due to pollution.

Local and New Zealand concerns:

Kind of. There are sometimes people coming in and cutting down trees in our neighbourhood. But, that's only like once every 3 years!

Definitely. Political debates and pollution – sometimes makes it hard to keep the environment safe.

Yes. I have 2 cats and they are amazing, but they sometimes catch birds. There are also a lot of possums around my house. Although my cats kill birds, they also kill other pests like rats, mice, and even a baby possum once.

Ravensdown "Gasworks station" [a fertiliser plant].

Yes, the sand that comes in from dredging the harbour.

New housing developments – are near my house, going through farmland and native bush.

Introduced species eating birds.

Some children talked about how they interact with nature in ways that are beneficial. What became obvious here was the importance of the family's own green space and the caring responsibilities it entailed. There was also recognition of the importance of local patches of native habitats.

Yes, humans can't survive without nature. I garden and walk dogs.

Nature does matter 2 our family. I have my own garden. We have sheep. Dad plants natives.

I think it's good to keep the world healthy and to keep NZ beautiful. We have a herb garden.

We use recycling bins and rubbish bins, throw out our rubbish. We also have lots of trees and gardens and everyone does lots of gardening.

We have lots of paddocks next to our property. There are lots of tree[s] and plants everywhere.

There is a lot of nature in our area and we have a electric car.

Yes! I have a forest really close to my house. I find it great to go on walks and simply admire nature. It is amazing to care for pets – even if they are as small as tadpoles.

There are a lot of trees behind our house and a lot of people have really nice gardens.

There are a lot of foresty areas in my neighbourhood but there are also a lot of possums.

Nature matters to my family because we have a farm.

We also (sometimes) pick up things that people dump on the side of the road because they think nobody will notice if they dump it on a gravel road.

On Gardens:

Lots of gardening happens and everyone is always out in their gardens.

We planted some flax, and we have a flower garden.

We do lots gardening and look after our trees and grass.

My mother is vegan and we are on a serious self-sufficiency year at the moment.

We have a garden that is terraced and grow some of our own food.

Because it is filled with natives and flowers.

We have a lot of garden. I walk outside and I'm practically attacked by all of our tomato plants [picture of "evil carnivorous tomatoes" included].

Yes, it's full of plants – some native – and plenty of birds come around. It's very green and there are some plants that are quite tall.

The questions to which children wanted answers

In the final part of the work with the older children, we adopted a different approach and asked the children to write what they thought were the important questions about nature. These questions ranged quite widely, from specific questions about matters such as photosynthesis and eating meat to deep questions about natural values, the people/nature relationship and pests. However, the list also includes some deeply ethical and challenging questions about climate change, extinction and nature's own development and relationships with people.

What benefits does nature have for your community?

What aspects of nature should we treat better? Why?

Do you think not eating meat (using another alternative instead) could help nature? If so, what would you suggest?

Are pests nature?

Why is nature important?

Are people defined as nature? And what would happen to nature if all people went "poof"?

Why do we feel protective of nature and endangered species?

Why does extinction matter when we kill millions of animals daily for food?

How is nature developing around the world?

Is nature destroying itself?

How does nature benefit us?

Why do some people want some animals and plant and tree species extinct?
What disbenefits does nature give us?
How old is nature?
How was photosynthesis figured out?
Is nature going to be perished in a matter of time? (plus picture of 'creepy smile')
Do you enjoy nature?
If most people know about climate change why doesn't anyone do anything about it?

In this selected sample, we see children from as young as six being able to relate to nature and also being curious about it, especially as it relates to nature in their own backyards. The older children still retain these characteristics but demonstrate in some cases quite sophisticated concerns about nature, especially at a global scale. They clearly relate to nature in an emotionally sensitive way, using words such as *peaceful, fun, beautiful, cool, interesting* and *quiet* when asked to describe nature. They also demonstrate a lack of value judgements in ways that are very different to adults. Particularly for younger children, pets are discussed simultaneously with wild animals, and flowers are admired for their own intrinsic properties – pretty and colourful. Only later do value judgements become more obvious in references to pests, native plants and with greater reference to native habitats such as bush and forest.

The children showed their sense of belonging through preferences for iconic native New Zealand species, even though introduced species may be more common. Although few children would have seen kiwi, they were very aware of their specialness and the need to protect them. They also made more frequent mention of native birds, such as fantails and tui. In Wellington, children mentioned the kaka (a threatened parrot),[5] as in this example: *Kaka come [from] Zealandia, and tuis that like eating the flowers that come off the flax*. In this quote, the child reveals a number of insights: recognition of the role of Zealandia, an urban wildlife sanctuary, in enhancing local bird populations, particularly that of kaka; the importance of kaka, and the relationships between native birds and native plants. In a study with children on Andros Island, where native species, as in New Zealand, are being displaced by invasive non-native species, researchers found the children exhibited an overall preference for the invasive species (Shapiro et al., 2017). The researchers argued that this preference will present difficulties for long-term species conservation. However, with reference to native species, for both the Andros and a comparative group of North Carolina children, the preference was for locally abundant or charismatic native species. Fortunately, the New Zealand children show no preference regarding common invasive species, such as rats, possums or stoats, and they do show preference for native charismatic species, such as fantail, tui, and kiwi, suggesting a more positive long-term conservation outlook than that for Andros. However, the issue of charisma is important, as non-charismatic species also have value and a role in ecosystem functioning, which can be overlooked and undervalued.

Where do children encounter nature[6] and influencing factors?

To encounter nature, two requirements are necessary. Firstly, there needs to be biodiverse spaces available, and secondly, children must be able to access these spaces. Not all spaces are accessible, and children's likelihood of accessing places varies. There has been much research on why children access some spaces and not others; here, we identify just some of the primary factors influencing access to natural spaces. Children often seek out spaces that are natural and show preference for outside spaces rather than indoor spaces. A school-based study in Iceland found children preferred playing outside and "favoured natural phenomena in the outdoor environment such as plants and animals, as well as variation in the landscape" (Norðdahl and Einarsdóttir, 2015, p. 158). They also showed strong preference for drainage ditches, even though these could be risky environments. The researchers noted that the children sought physical challenges and mostly preferred to be out with other children. This preference for natural/informal spaces is a common finding from diverse environments, including Scottish city and rural environments (Day and Wager, 2010) and urban Korea (Min and Lee, 2006). Three other key factors emerge in supporting accessing of space: a sense of belonging, safety and companionship.

The prerequisites for nature encounter: belonging, safety and companionship

The role of territoriality and belonging is an important foundation for building nature connection. Children tend to gravitate out from their central space of belonging. The central space is one where they form attachments based on favourite activities and which affords a sense of security (Chatterjee, 2005). A sense of safety is fundamental. Children are very aware of and will not use places where they feel threatened by other users, including other children. Interviews with children in South Africa detected a number of common themes emerging in terms of children's fear of natural places, a fear they termed a fear commonly referred to as: "nature as the dangerous other" (Adams and Savahl, 2015). This fear was also reflected in children living in the more deprived urban neighbourhoods in the Scottish study, where it acted to inhibit their use of natural spaces such as parks (Day and Wager, 2010). As Nordström neatly puts it, following her work with Finnish children, "Safety has a fundamental value for children and only when that has been satisfied does physical environment play an emotionally important role" (2010, p. 523). This feeling of security is an important determinant of where children play and explains the importance of the immediate, places close to home. These immediate spaces include semi-private threshold places, the liminal spaces between home and the street (Carroll et al., 2015, Figure 2.4), driveways, verges, footpaths and local communal spaces, which can include school grounds, local parks and sports fields as well as the 'left over' bits of land that no one really claims and on

Figure 2.4 A threshold space (between house and pavement) in Amsterdam being used as a place for nature encounter.

which nature often makes its mark. These spaces can often provide the substance of nature encounters for children in their daily lives.

Related to the question of security and nature connection is that of social connection. In our *Natural Neighbourhoods for City Children* study, we looked at social connections by asking children on their computer-drawn maps to identify the homes of their friends and people they knew. In common with an earlier study on social connections (Freeman, 2010), the children showed very different levels of social connection, from knowing no one to being able to name or recognise most of their immediate neighbours and many further away. We know from research by Mikkelsen and Christensen that "children's mobility has to be recognised as primarily social, and that companionship pervades every aspect of the activity" (2009, p. 37). Sociability, in particular having friends, is a prerequisite for outdoor play and for play in the wider neighbourhood (Figure 2.5).

The importance of social relationships in nature encounters was evident when we asked children about their preferred place to play when alone and with friends. On their own, children primarily play inside or in their own yard/garden. Some do go to the park but not as many as when they play with friends (Figure 2.6). When

Figure 2.5 The evident sociability of childhood in Cape Town means that children are seen out and about. Safety is further enabled through the presence of local residents undertaking everyday activities, such as fishing, and there is a communal responsibility for children.

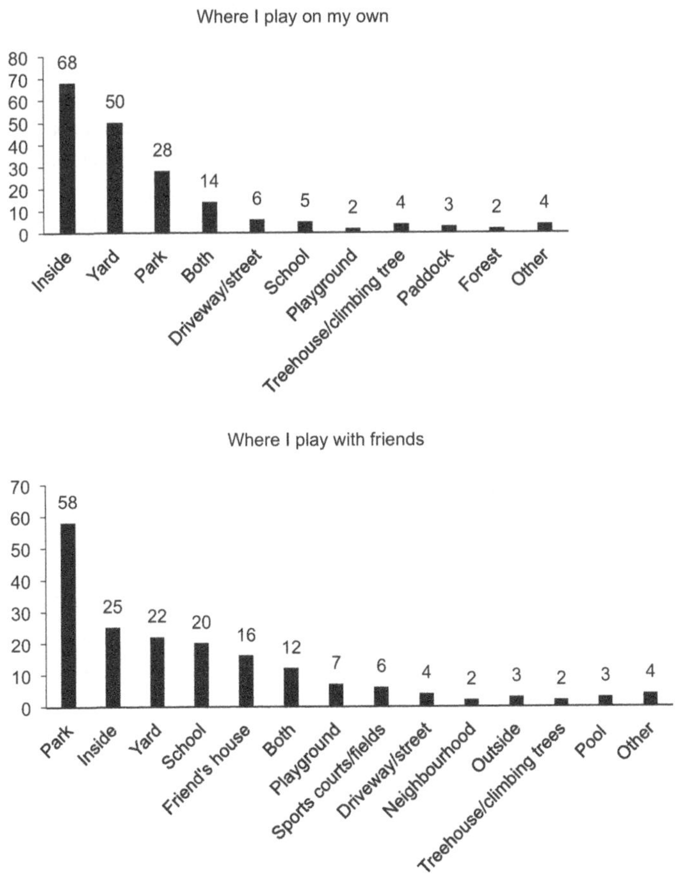

Figure 2.6 Where children play when alone and with friends; findings from the *Natural Neighbourhoods for City Children* study.

with friends, the children were much more likely to refer to outside locations and referred to a wider range of locations. Some played inside with friends, but there were only 25 references from 187 children so that was quite low. This preference for being inside when alone and outside when with friends applied to children across low, medium and high socio-economic groups. Social connections thus enable children's encounters with outdoor and natural spaces, and it is to the spaces of encounter that our attention now turns.

Where do children go and why?

In the *Natural Neighbourhoods for City Children* study, we were interested to know where children go to encounter nature and why. This was not a novel question, but the way children provided the information was, through the use of personalised computer nature maps. Various methods have been used to identify children's journeys and the places they visit, including drawings and hand-drawn maps (see Table 2.1 and Figures 6.2a and 6.2b), GPS monitors that the children wear, travel diaries, recall during interviews and point information inserted onto maps through use of a GIS. In our study, we used a number of methods. These were a hand-drawn map showing nature in their home area, an interview where they talked about where they went and most importantly a personalised computer nature map where children placed dots on a map showing places they use, with the number of dots indicating the amount of time spent in each place (Freeman et al., 2016). In addition, children indicated places on their map with nature values, where they would see bugs, birds, natural water and special trees and experience a quiet nature place or a wild nature place. Also included were social and activity features, such as their school, friends' and neighbours' houses, parks and other features of interest. As they added information to their maps, they talked about what they were doing and why. For example, they could say *I see birds on the power lines, in the trees behind my house, bugs on the deck, I go to the field to play football*, or *that's where we grow vegetables*. These personalised computer maps enabled the researchers to get a good idea of which natural spaces children used, how often and the reasons why they used them.

Distances travelled and home ranges: The children rarely travelled independently beyond their neighbourhood, and for the few children who did, it tended to be a predetermined route to a specific destination, such as to a relative's house. Only 4% of children travelled freely around the suburb and were allowed beyond the suburb. No child had free access even to areas adjoining their suburb. Some 22% effectively could not go very far from their home and had their freedom limited to their garden and their street. Some 30% were allowed to go further but only for a specified journey in the neighbourhood, usually to the school. When we examined more closely the spaces children could access, the limitations placed on children in accessing natural spaces became more apparent (Table 2.3).

The median independent home range size was approximately 6 ha, but when we removed all the areas that are not accessible to children, as they are private

Table 2.3 Levels of independence for children in the *Natural Neighbourhoods for City Children* study

Levels of independence	*% of 187 children*
No independence	10
Home/surrounds/street only	12
Home surrounds/street and one other place (e.g., school)	22
Home surrounds/street and freedom within the suburb but limited to specified journeys/destinations	30
Home surrounds/street and freedom within the suburb, not limited	22
Freedom within the suburb and specified destinations outside suburb	4

or inaccessible land (e.g., someone's garden), the remaining area that was accessible was 2.4 ha. Thus, approximately half their home range area was inaccessible. The average maximum distance from home for children was about 500 m with a few exceptions for some children who went on longer trips to specified destinations. Most children had parental restrictions placed on where they could go. Parental restrictions were similar across all demographic groups. The amount of biodiversity available to children that they could access independently was driven most by individual and social factors, in that these acted to constrain the size of the child's home range. Within the home range, the most important factor determining the biodiversity children might encounter was the biodiversity of accessible gardens. Because children's home ranges often lacked the most biodiverse and natural habitats present in their neighbourhood, gardens were left as the most biodiverse accessible habitat. Given that private gardens often comprised a large proportion of the home range, particularly for those children with a small home range, these areas strongly influence children's time in, and experience of, nature.

Natural areas available and accessed: Over 50% of the neighbourhood area was made up of private gardens, with green spaces making up another third. These green spaces included sports fields, school grounds, cemeteries, golf courses, parks and other incidental green spaces. The majority of children also had easy access to woodland, vacant land and recreational green habitats close to home. Nearly all children had an accessible garden and 90% a park that they could visit close to their home. The areas that children most used were gardens and recreational green spaces (playgrounds, parks, sports fields). Fewer children accessed the more biodiverse spaces, such as woodland/forest, streams and beaches, although these were often available. We developed a method to quantify the biodiversity value of each child's home range (Hand et al., 2016). Garden and neighbourhood biodiversity was linked to socio-economic status, but the difference was primarily related to the greater cover of mature gardens of high biodiversity value in areas of higher socio-economic status, with more deprived neighbourhoods containing lower biodiversity gardens. The biodiversity value and number of public green spaces were very similar for children

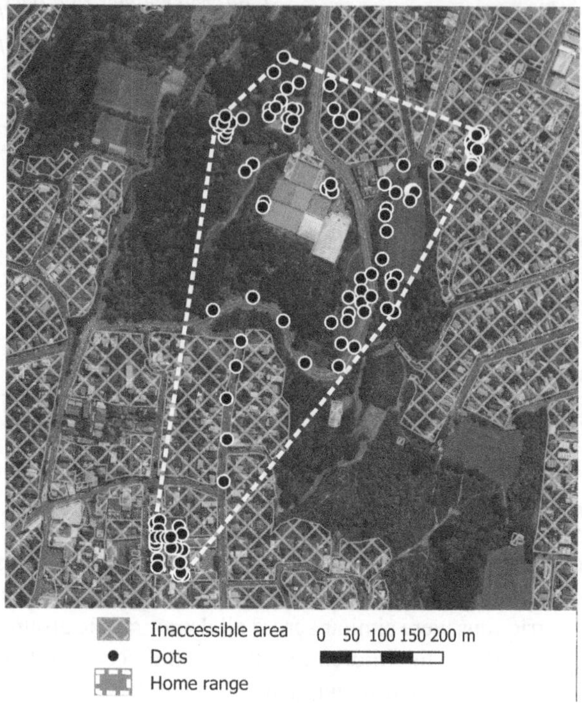

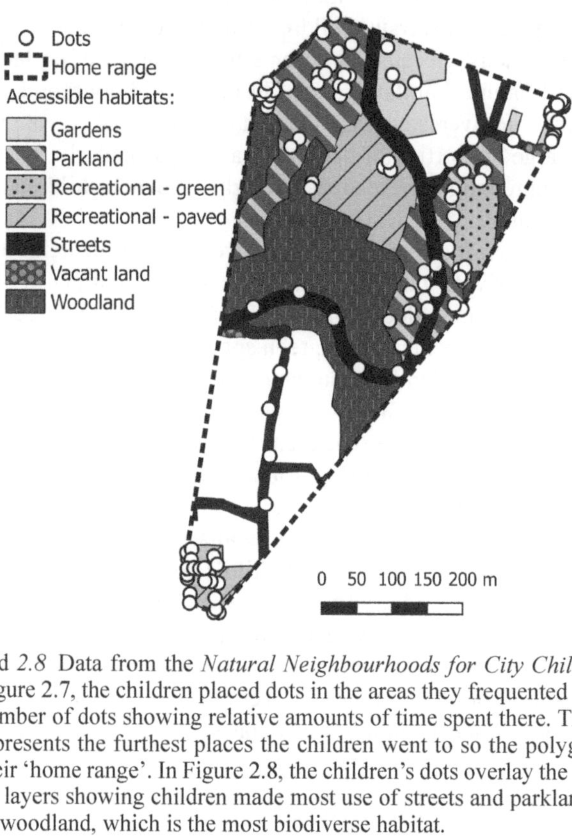

Figure 2.7 and *2.8* Data from the *Natural Neighbourhoods for City Children* study. In Figure 2.7, the children placed dots in the areas they frequented most, with the number of dots showing relative amounts of time spent there. The dashed line represents the furthest places the children went to so the polygon represents their 'home range'. In Figure 2.8, the children's dots overlay the different habitat layers showing children made most use of streets and parkland with no use of woodland, which is the most biodiverse habitat.

irrespective of socio-economic status. Children in the study did not access the most biodiverse spaces in proportion to their availability (Figures 2.7, 2.8). As gardens were found to be the spaces where many children spend most (>40%) of their time outdoors (and for children with the most restricted independence, their only accessible outdoor space), children living in more deprived neighbourhoods are at a disadvantage in experiencing nature.

Children's understanding and local nature: While biodiverse spaces were apparently available to all children in the study, given that all neighbourhoods had a range of biodiverse spaces, in reality, many of the children did not use them. Despite this, nearly all children did have spaces where they could access nature. Even the small number without garden access had locally available biodiverse spaces that could be accessed with family or friends. All the schools had biodiverse spaces, including trees and grassed areas; some had vegetable gardens and all had landscaped areas. Some had fields/paddocks or forest/bush/woodland adjacent or nearby rather than on site.

We calculated a nature knowledge score for each child based on what he or she knew about nature in his or her neighbourhood (Freeman et al., 2017). This score was calculated from the information they added to their personalised computer nature map, such as where birds, natural water, a wild animal and insects could be found or where a quiet nature place or a 'wild jungly' type space or a special tree might be found. The scores were calculated from responses to seven questions, with each response ranked between 0–3, giving a potential total of 21 if the child showed 'full' nature knowledge. The majority (58%, n = 109) of the children's nature knowledge scores fell between 11 and 20, indicating reasonably good knowledge of nature within their neighbourhoods; fewer than 10% had very limited or no knowledge (Freeman et al., 2017). Though most children knew where to go to see animals, such as birds or insects, fewer were able to make clear linkages between animals and their habitats. We further investigated children's scores to identify the factors that might explain variation in individual scores (Table 2.3). One of the variables we were interested in was independence, to see if there was a relationship between nature knowledge and where children go without adults. Another variable was ethnicity where we found that ethnicity and city had the greatest influence on nature connection, with children in some cities scoring higher than in others and children of Asian origin scoring lowest. Within each ethnic group, the relative importance of social and environmental factors differed. For Māori, Pasifika and children of Asian origin, social factors such as independence, socio-economic status and the number of social connections in the neighbourhood were associated positively with nature knowledge, whereas among children of European origin, higher nature scores were positively associated with the biodiversity values of their immediate neighbourhoods. These findings emphasise the role played by both social and environmental factors, and the importance of gardens as accessible biodiverse spaces in children's home ranges (Figures 2.9a, 2.9b and 2.9c).

Figure 2.9 A number of children mentioned their own gardens when talking about responding to nature's challenges. Most of the children will have their own garden, and gardening is an activity that most New Zealand children engage in as in these photos of making compost and digging. Several of the younger children in the study mentioned going on the trampoline in the backyard to be in nature.

Re-establishing nature access

Given that children all had restrictions on which spaces they could access, which for some was only their garden, these results show that even though the current generation of children are unable to access spaces available to children in previous generations, they are not necessarily disconnected from nature. Children clearly do make connections within the areas available to them, even where this area is very small. The challenge is to give all children access to biodiverse spaces, which can be immediately around their homes, but also to reinstate children's access to the biodiverse spaces that are available. The children in our study lived in cities that are by most standards high in biodiversity, yet few children were able to freely access that biodiversity.

Children's use of outdoor spaces is dependent on these spaces being safe and inviting. They provide the challenges and freedom children desire, a freedom that is not possible when under adult supervision and control (Figure 2.10). In planning and managing urban environments, children under the UNCRoC have the right to have their say on their environments. Our research has shown that children can provide insights into their environment that are valuable to those whose professional remit determines the shape of that environment. Given access

Figure 2.10 Natural spaces can provide social refuges for children and young people away from adults. Young people have created this this 'bush' camp in Brisbane, Australia, with its own sitting place and skateboard ramp. In this setting, incidental encounters with snakes, wallabies, parrots and other species are highly likely.

Figure 2.11 In this photo, we see an attempt to reinstate children's right to roam.

to appropriate tools, they can plainly demonstrate their understanding and knowledge of that environment. However, it also has become clear that children, not just those in our New Zealand study, are being constrained in their environmental access. Their freedom to use the most biodiverse spaces is declining as parents across a wide spectrum of urban environments curtail their children's access, in large part because of concerns for their safety and to avoid risk and potential harm (Figure 2.11). Natural environments do contain risk, but if children do not learn how to engage with them, then we would argue that in the long term, the risk is greater. Lack of association means they lose access to the benefits nature offers and they are less likely to develop lifelong concern for nature and its protection (Wells and Lekies, 2006). Where access was limited, the children in our study did seek out nature; even where their freedom was severely curtailed, they still connect with nature, albeit in limited form through, for example, being able to know where to find bugs or birds or identifying special trees. The older children were able to clearly articulate their concerns for the natural world and possessed valuable insights on the changes necessary to address these. Although children seem less able to freely connect with wild spaces, we see their voices as giving us hope that they value nature, seek out natural connections and are interested in the natural world in various forms in the places where they live.

Notes

1 We also asked this question in a study with older adults, and they also found it difficult and their answers were not always dissimilar to those of the children.

2 The quotes from the 6-year-olds and from the 10–11-year-old children from the *Natural Neighbourhoods for City Children* study were transcribed by the interviewer. The 12–13-year-olds wrote their own quotes, but we have corrected the spellings in some cases.

3 The interviewer transcribed the quotes from the 6-year-olds and from the 9–11-year-old children in the *Natural Neighbourhoods for City Children* study. The 12–13 year olds wrote their own quotes but we have corrected the spellings in some cases.

4 In New Zealand, there has been a lot of debate about a methane tax possibly being levied on dairy farmers, as the cattle are responsible for most of New Zealand's methane emissions. This accounts for a number of mentions of cows and methane.

5 A reintroduced native parrot that has successfully bred in the Wellington Eco-sanctuary Zealandia, subsequently radiating out from the sanctuary across the urban area as the population grew; now some 350–400 kaka live in Wellington.

6 The data in this section are taken from the following papers and two others in press at the time of writing.

Hand, K.L., Freeman, C., Seddon, P.J., Stein, A., and Van Heezik, Y., 2016. A novel method for fine-scale biodiversity assessment and prediction across diverse urban landscapes reveals social deprivation-related inequalities in private, not public spaces. *Landscape and Urban Planning*, *151*, pp. 33–44.

Freeman, C., van Heezik, Y., Stein, A., and Hand, K., 2016. Technological inroads into understanding city children's natural life-worlds. *Children's Geographies*, *14*(2), pp. 158–174.

Freeman, C., van Heezik, Y., Hand, K., and Stein, A., 2015. Making cities more child- and nature-friendly: A child-focused study of nature connectedness in New Zealand cities. *Children, Youth and Environments*, *25*(2), pp. 176–207.

References

Adams, S., and Savahl, S. 2015. Children's perceptions of the natural environment: A South African perspective. *Children's Geographies* 13(2): 196–211.

Carroll, P., Witten, K., Kearns, R., and Donovan, P. 2015. Kids in the city: Children's use and experiences of urban neighbourhoods in Auckland, New Zealand. *Journal of Urban Design* 20(4): 417–436.

Chatterjee, S. 2005. Children's friendship with place: A conceptual inquiry. *Children Youth and Environments* 15(1): 1–26.

Childwise UK. 2016. *Childwise Monitor Report 2016, Connected Kids*, Childwise, Norwich UK.

Clarkson, B.D., Wehi, P.M., and Brabyn, L.K. 2007. A spatial analysis of indigenous cover patterns and implications for ecological restoration in urban centres, New Zealand. *Urban Ecosystems* 10(4): 441–457.

Day, R., and Wager, F. 2010. Parks, streets and 'just empty space': The local environmental experiences of children and young people in a Scottish study. *Local Environment* 15(6): 509–523.

Freeman, C. 2010. Children's neighbourhoods, social centres to 'terra incognita'. *Children's Geographies* 8(2): 157–176.

Freeman, C., and Buck, O. 2003. Development of an ecological mapping methodology for urban areas in New Zealand. *Landscape and Urban Planning* 63(3): 161–173.

Freeman, C., Stein, A., Hand, K., and van Heezik, Y. 2017. City children's nature knowledge and contact: It is not just about biodiversity provision. *Environment and Behaviour*, Advance online. http://dx.doi.org/10.1177/0013916517732108

Freeman, C., van Heezik, Y., Hand, K., and Stein, A. 2015. Making cities more child-and nature-friendly: A child-focused study of nature connectedness in New Zealand cities. *Children, Youth and Environments* 25(2): 176–207.

Freeman, C., van Heezik, Y., Stein, A., and Hand, K. 2016. Technological inroads into understanding city children's natural life-worlds. *Children's Geographies* 14(2): 158–174.

Freeman, C., and Vass, E. 2010. Planning, maps, and children's lives: A cautionary tale. *Planning Theory & Practice* 11(1): 65–88.

Hand, K.L., Freeman, C., Seddon, P.J., Stein, A., and Van Heezik, Y. 2016. A novel method for fine-scale biodiversity assessment and prediction across diverse urban landscapes reveals social deprivation-related inequalities in private, not public spaces. *Landscape and Urban Planning* 151: 33–44.

Hill, M., Davis, J., Prout, A., and Tisdall, K. 2004. Moving the participation agenda Forward. *Children & Society* 18: 77–96.

Holloway, S., and Valentine, G. 2003. *Cyberkids: Children in the Information Age*, Routledge Falmer, London.

Louv, R. 2005. *Last Child in the Woods: Saving Our Children from Nature-Deficit Disorder*, Chapel Hill, NC: Algonquin Books of Chapel Hill.

Mikkelsen, M.R., and Christensen, P. 2009. Is children's independent mobility really independent? A study of children's mobility combining ethnography and GPS/mobile phone technologies. *Mobilities* 4(1): 37–58.

Min, B., and Lee, J. 2006. Children's neighbourhood place as a psychological and behavioral domain. *Journal of Environmental Psychology* 26(1): 51–71.

New South Wales Commission for Children and Young People. 2009. *Built for Kids: Child Friendly Communities*, NSW Commission for Children, Sydney, Australia.

Nordström, M. 2010. Children's views on child-friendly environments in different geographical, cultural and social neighbourhoods. *Urban Studies* 47(3): 514–528.

Norðdahl, K., and Einarsdóttir, J. 2015. Children's views and preferences regarding their outdoor environment. *Journal of Adventure Education & Outdoor Learning* 15(2): 152–167.

Of Com: UK. www.ofcom.org.uk/ac.

Shapiro, H.G., Peterson, M.N., Stevenson, K.T., Frew, K.N. and Langerhans, R.B., 2017. Wildlife species preferences differ among children in continental and island locations. *Environmental Conservation, 44*(4), pp. 389–396.

UN General Assembly. *United Nations Convention on the Rights of the Child.* www.refworld.org/docid/3ae6b38f0.html

Wells, N.M., and Lekies, K.S. 2006. Nature and the life course: Pathways from childhood nature experiences to adult environmentalism. *Children Youth and Environments* 16(1): 1–24.

Part II

Understanding urban biodiversity in a changing environment

3 Wild nature, biodiversity, habitats and species

An ecological approach

Ecologists see urban areas differently from most of the people who live in them. Towns and cities are environments that have been built specifically to cater to the needs of our own species, and once built, cities usually remain in place, with very little chance of any future regeneration back to a more natural state. Despite our single-minded focus on our own needs, a number of other species are able to exist across this built environment, and the existing ecosystems provide us with a range of services that most people are unaware of and take for granted. An ecologist sees the urban landscape as a complex mosaic of habitats, ranging along a spectrum of modification, from the highly modified central business district and areas supporting heavy industry, to suburbs, which are also modified, but generally much greener. The biodiversity and ecosystem values of these habitats are affected by the degree of modification, their structure and their connectedness. It is very clear that urbanisation has altered hugely plant and animal species diversity, distributions, and abundance: some species have adapted, some have even benefited, while other less-tolerant species have disappeared. In some cases, urban areas can act as refuges for species that are steadily disappearing from modified rural landscapes (Goddard et al., 2010). In this chapter, we describe the urban environment in the context of spaces where other species live and the characteristics of species that are able to tolerate and even thrive in this human-dominated environment. We explore the potential that different typical modified urban habitats have to support biodiversity, as well as the ways that animals have responded to the challenges posed by urban living. Knowing more about the urban habitats and biodiversity present in urban environments is an essential prerequisite to understanding the opportunities available to children living in cities for experiencing and connecting with nature.

Characteristics of urban habitats

Urbanisation is the transformation of natural environments into built environments. This transformation can be extreme, where all natural habitat is lost, such as commonly occurs in the centre of the city. It can also be less extreme, where natural habitat has been replaced by a mixture of native and non-native vegetation in low-density built environments interspersed with patches of natural habitat, such as in the suburbs. On the edges of the city, larger patches of natural habitat may continue to co-exist with agricultural land. The three main properties that distinguish urban landscapes from natural ones are the changes to the abiotic environment that have occurred, the organisation and connectivity between patches of habitat and the animal and plant communities they support and the high number of invasive species (Niemelä et al., 2009).

Biogeochemical cycles are altered in cities: industry and transportation contribute CO_2 and other greenhouse gases, which affect climate, and air pollution influences nutrient cycling and primary production, contributing to eutrophication in rivers and coastal waters (summarised in Grimm, 2008). High concentrations of chemicals in the environment and altered hydrography have both growth-enhancing and growth-inhibiting effects on plants. Local changes in climate produce an urban heat island effect, such that cities tend to have higher air and surface temperatures than the surrounding rural areas. Average temperatures in cities can be 1 to 2 degrees Celsius higher during winter and 0.5 to 1.0 degree Celsius higher during summer in the mid-latitudinal US (Botkin and Beveridge, 1997), and there is less thermal variability. Higher temperatures affect biodiversity and ecosystem functioning, as well as water resources, air quality and human health (Grimm, 2008). For example, growing seasons are longer in urban areas, which affects the amount and timing of availability of food resources for wildlife.

The urban landscape is made up of a large number of small patches of different habitats that vary in their capacity to support wildlife populations, depending on the extent to which they have been modified. The opportunities children have to encounter wildlife will depend on the ease with which they can access these habitats and the types of animals occupying them. Fragments of natural vegetation, vacant lots, and riparian strips are likely to be the most natural. These patches are usually isolated from each other within a matrix of built and modified environments, making dispersal for some organisms between patches highly risky. Connectivity between patches is important for species persistence if the small size of the patches means they support only small populations of animals. The configuration, size and degree of connectivity between patches will have a major impact on the ease with which populations in patches are able to mix. Connectivity can be enhanced by green corridors or by well-vegetated matrix habitat, such as urban gardens. However, the harshness of the matrix can vary hugely, with areas almost devoid of vegetation acting as barriers to animal movements. Those species that have wide habitat requirements are able to disperse across landscape gaps, and through modified habitats, and are more likely to thrive in urban environments.

Urban areas support many non-native (exotic) species. Because people like colour and variety in their parks and gardens, exotic plants often outnumber natives across city neighbourhoods. In the New York metropolitan area, the number of native woody plant species (species richness) declined along the gradient of increasing urbanisation, while exotic woody plant richness increased (Aronson et al., 2015). In New Zealand, private gardens typically contain twice as many exotic as native woody plant species, and many gardens have no native woody plants at all (van Heezik et al., 2014). A slightly higher proportion of native species was recorded in UK gardens, where 30% of the garden flora and about one-third of the 50 most frequently occurring species were native, but the majority of garden plants are exotic (Knapp et al., 2012; Smith et al., 2006; Thompson et al., 2003).

The replacement of regionally distinct plant and animal communities and endemic species with a few widespread exotic species results in city faunas that can be remarkably similar in many different parts of the world, with common species such as house sparrows (*Passer domesticus*), starlings (*Sturnus vulgaris*), and rock doves (*Columba livia*) (McKinney and Lockwood, 1999; Clergeau et al., 2006). The proportion of exotic birds in urban communities varies hugely across cities and countries; surveys in Sheffield in the UK revealed only three non-native species, whereas in Dunedin, New Zealand, nearly half of all species surveyed were non-native (Fuller et al., 2009; van Heezik et al., 2008). Non-native species are usually most abundant at the extremes of the urban habitat gradient, where only a small number of species can thrive. While some believe that exotic birds compete with native birds for resources and disperse exotic weeds, very little evidence has been collected to support negative interactions such as these. Many exotic birds are valued by adults and children for their colours, song and ease of viewing. While the quality of the nature experience for children is unlikely to be affected by the status of the species, an awareness of which species are native may promote a sense of identity or place.

Species occurrence and distributions

The high level of patchiness of land covers across urban landscapes results in large numbers of species (species richness) overall. However, diversity of animals declines across the urban gradient into the city centre, where large numbers of a restricted range of species are usually found (e.g., rock doves or house sparrows; McKinney, 2008; Faeth et al., 2011), and many species that are specialised to particular habitats are absent. Diversity at moderate levels of urbanisation, i.e., suburban areas, varies among different taxa. In the case of birds, suburbs provide habitats which some native birds are able to use as well as non-natives, resulting in high species richness (van Rensburg et al., 2009; van Heezik et al., 2008), but the same trend is not often observed for invertebrates and non-avian species (McKinney, 2008). Patterns of diversity in relation to urbanisation also vary between cities. In some cities, overall species richness does not change if native species that are lost are replaced by exotic species (Aronson et al., 2015). Species diversity is important, since the psychological benefits enjoyed by people from their contact with nature increase with the number of species people perceive in urban green spaces (Fuller et al., 2007a).

Species diversity is very much influenced by the mixture, quality and configuration of available habitats (van Heezik and Adams, 2014; Jokimäki and Suhonen, 1998), and for this reason, diversity is highest in patches of near-natural vegetation, such as woodlands and river corridors (Stagoll et al., 2010). The structure of the habitat, the plant species and the number of layers of vegetation and plant forms are the main factors influencing community diversity. For example, in Australia, Threlfall et al. (2016) found that increasing amounts of native vegetation, understorey vegetation volume, and large trees were all effective in increasing the species richness of bats and birds. Ultimately, habitat quality influences how much food is available to wildlife, the level of disturbance and the existence of nest sites and refuges. Large expanses of good habitat are rare, and it is more common that patches of natural or semi-natural vegetation are embedded in a matrix of unsuitable sites.

In the case of predators, some exist at higher densities in urban areas (Sorace, 2002). Urban carnivore distributions are determined by local factors such as the composition of the vegetation, habitat complexity, the presence of water and the density of prey and other predator species, but also by socio-economic factors for species that rely on human food sources (Magle et al., 2015). In Chicago, for example, coyote (*Canis latrans*) populations are most likely to survive within wealthier neighbourhoods. Some predators avoid urban areas, whereas others, such as racoons (*Procyon lotor*), cats (*Felis catus*) and corvids (crow family), have become common. Some evidence points to urban habitats being safer from predation than rural and natural habitats (Shochat et al., 2006), despite very high cat densities. It is possible that the species persisting in cities are to some extent 'cat resistant'.

Less is known about urban invertebrate communities, but like birds, invertebrates respond negatively to habitat loss and fragmentation and to changes in the abiotic environment, pollution, high levels of human disturbance and loss of host plants (Jones and Leather, 2012; McDonnell and Hahs, 2008; McKinney, 2008). Communities of arthropods in urban areas are dominated by a few synanthropic species, i.e., wild animals and plants of various kinds that live near, and benefit from, an association with humans and the somewhat artificial habitats that humans create around them that become very abundant (McIntyre et al., 2009). This includes species such as Argentine ants (*Iridomyrmex humilis*) and European cabbage butterflies (*Pieris rapae*) (reviewed in McIntyre and Rango, 2009). The spread of species such as these that are able to exploit the urban environment worldwide has resulted in homogenisation of the urban arthropod fauna. Beetles (Insecta, Coleoptera) have been the focus of most of the urban invertebrate research. Beetles tend to decline in diversity across the gradient of increasing urbanisation, as forest specialist species are lost (see van Heezik et al., 2016). However, the variety of open habitats, such as lawns, bare soil, moss and gravel results in an increase in open-habitat specialist, generalist and opportunist species (Gagné and Fahrig, 2011; Jones and Leather, 2012; Madre et al., 2013; Tóthmérész et al., 2011; Vergnes et al., 2014). Like birds, beetle species richness and abundance are related to the amount and structure of vegetation, such as the number of trees, the canopy area and the number of native plants, as well as how well their habitat patches are connected to other woody habitats (Smith et al., 2006a, 2006b; van Heezik et al., 2016; Vergnes et al., 2012).

While some studies report that native invertebrate species can be negatively affected in environments dominated by exotic plants, this is not the case for all

Figure 3.1 Huhu beetles are not often seen in urban habitats. This one was rescued from drowning in a dog's water bowl and provided an unusual opportunity to get close up to this large bug.

species (Burghardt et al., 2008; Derraik et al., 2005; Helden et al., 2012; Pardee and Philpott, 2014; Simao et al., 2010) (Figure 3.1). Burghardt et al. (2008) found fewer Lepidoptera associated with exotic plant-dominated gardens in the US, and in Singapore, cultivated greenery was associated with homogeneous bird and butterfly communities compared with more natural vegetation (Chong et al., 2014). In contrast, in California, native butterflies breed mostly or entirely on exotic plants (Shapiro, 2002), and in New Zealand, most of the beetles identified in private gardens were native despite there being a high proportion of exotic vegetation (van Heezik et al., 2016). This was unexpected, as invertebrates are thought to co-evolve on the plants they rely upon, so that they are unlikely to occur in the absence of the plant. In fact, New Zealand gardens had a greater diversity of beetles than has been found in other natural and modified habitats. However, habitats that urban areas often lack are the messy ones, with rotting wood from downed trees. These provide important habitat for some species, such as Onycophorans, which are associated with large decayed cover objects and downed woody material (Barrett et al., 2016). In general, the kinds of species children are likely to encounter in their neighbourhoods will depend on where they live within the city, but even in suburban neighbourhoods, the species they interact with are likely to be non-native generalists that tolerate human disturbance well.

Modified landforms and their role in maintaining biodiversity

Virtually all of the urban landscape is modified to some extent, but these modified landforms still play an important role in supporting populations of animals and facilitating dispersal. They are also places where children might play. There is huge potential to manage modified landforms in such a way as to achieve an optimal solution, integrating functionality with biodiversity values.

Parks

Parks are popular places where children play, as they can offer a range of opportunities for different types of play. Because urban parks typically contain a wide variety of land uses and habitats, they also have the potential to support many species; however, parks are often intensively managed (e.g., mowing, spraying of chemicals, pruning and removal of dead trees) and subject to a lot of human disturbance. Increasingly parks are relaxing their management regimes to encourage the development of more natural habitats (Figure 3.2). The location and size of parks will also have a significant impact on the wildlife they

Figure 3.2 The trend is towards more natural landscapes in parks as shown in this photo of Dunedin Botanic Gardens, New Zealand. Here, the grass has been left uncut, providing better wildlife habitat and a more natural play experience.

support. When parks are situated within high-density developments and in the commercial core of the city, human activity is high and bird species richness is low, often with a bird community dominated by exotic urbanophilic species. Vegetation cover in high-density areas is likely to be lower, which has the effect of isolating the park from other areas of wildlife habitat; this will affect some species more than others.

Gardens/urban agriculture

Urban agriculture takes many forms, from private gardens to community or allotment gardens, rooftop farms and urban orchards. Although managed intensively, urban agriculture can maintain high levels of biodiversity. Over a thousand plant species have been recorded from 267 gardens across five cities in the UK, more than in other local urban and semi-natural habitats (Loram et al., 2008). In Stockholm, where allotment and community gardens are often large and well-established, a single $400m^2$ garden can contain more than 440 plant species; these support urban pollinators for long periods of time because of the rich abundance of flowering plants (Colding et al., 2006). In Brazil, 98 plant species were recorded from 21 urban gardens, including a large diversity of fruit trees and shrubs, as well as vegetables, herbs, and medicinal plants (WinklerPrins, 2002).

Most plant species in gardens are non-native. In Hong Kong, 82% of tree species were exotic, and they made up 92% of the total tree count (Zhang and Jim, 2014), while in New Zealand, only 17% of the 553 species recorded across 55 gardens were native (van Heezik et al., 2014). Some non-native species do provide resources for native wildlife such as pollinators and birds (Shapiro, 2002; French et al., 2005), and exotic birds can contribute to dispersal and pollination (Kelly et al., 2010). In New York City, community gardens dominated by exotic plants supported 54 bee species (Matteson et al., 2008). In Dunedin, New Zealand, despite a very high proportion of exotic woody plants, the invertebrate community, particularly the beetles, was predominantly native and surprisingly species-rich, more so than in other less-modified habitats (Barratt et al., 2015; van Heezik et al., 2016). However, the intensive management regimes in gardens (mowing, pruning, pesticide application) limits their capacity to maintain populations of rare and sensitive species (Matteson and Langellotto, 2010) (Figure 3.3a and 3.3b).

Golf courses

Golf courses may seem unlikely places for children, but their edges can offer interesting spaces for play. Golf courses cover large areas, 70% of which are non-playable (for golfers) areas of streambanks, wetlands, grasslands, groups of trees, and woodlands. Colding et al. (2006) estimated that around 40 ha of a typical Stockholm golf course is made of natural habitats, an area similar to that of median-sized Stockholm nature reserves. Golf courses provide wetland habitat, which is typically absent in urban areas. When appropriately managed and designed, they have

a

b

Figure 3.3a and *3.3b* Contrasting small gardens, one designed to support natural features, the other to remove them. The opportunities for children to experience connection to nature are equally contrasting.

the potential to support reasonably abundant wildlife, including species that are not found in the surrounding suburbs. In California and Ohio, golf courses supported the greatest number of species along an urban gradient, probably because of the variety of habitats (Blair, 2004). In Queensland, Australia, the biodiversity value of golf courses varies, with some supporting high densities of regionally threatened vertebrates, while others support only common urban-adapted species (Hodgkison et al., 2006). In Australia, golf courses that contain largely native plants support a higher diversity of insectivorous birds and bats than can be found in surrounding residential areas. This is possibly due to a greater volume of understorey vegetation and higher proportion of native plants, as well as the large unlit areas that provide nocturnal species with foraging habitat free of artificial lighting (Threlfall et al., 2016).

Cemeteries

Again, cemeteries may not be perceived as typical areas that children might frequent, but it is unlikely that children located close to a cemetery do not use it in their play. The potential for urban cemeteries to support biodiversity depends on the cemetery's age and management regime (Figure 3.4). For example, 604 species (bats, birds, lichens, bryophytes, carabids, vascular plants, spiders) were

Figure 3.4 Kangaroos have become habituated to living alongside people in Pinnaroo Valley Memorial Park, Perth. The park maintains both lawns and natural bushland and markets itself as the most environmentally responsible cemetery in Australia.

recorded in an old cemetery in Berlin, including some species of conservation concern. Species richness and abundance were strongly dependent on the intensity of management, which determined the volume and structure of vegetation (Kowarik et al., 2016).

Characteristics of urban species – what makes them succeed or fail?

Among birds, which are the most intensively studied taxon in urban areas, native species respond to the process of urbanisation by separating into a handful of winners which are able to adapt to or exploit human-modified environments, and many losers, which tend to be local endemic species that are intolerant of human-dominated landscapes (McKinney and Lockwood, 1999). What characteristics allow populations of some species of birds to persist or even thrive, while others fail? Adapters are often more social, sedentary (as opposed to migratory) and able to exploit human-derived sources of food (Figure 3.5). Sociality may help urban exploiters to locate, communicate about and compete with other species over sources of food in a highly patchy environment and at the same time to avoid predators (Kark et al., 2007).

Figure 3.5 These South American seals, watched by local children, gather by the fish market at Punte del Este, Uruguay, to take advantage of the fish guts being thrown into the sea.

Foraging and nesting guilds

The foraging guild[1] of species, which is defined by the type of foods a species normally eats (e.g., seeds, fruit, invertebrates, nectar) often predicts winners and losers: urban adapters in temperate regions tend to be omnivorous (eating plants and animals) or granivorous (eating seeds) (Lim and Sodhi, 2004; Kark et al., 2007; Sol et al., 2002). Omnivores commonly exploit anthropogenic food sources (i.e., supplied intentionally or indirectly by people), which can help them to survive severe winters. The abundance of granivorous species, such as house sparrows (*Passer domesticus*), typically increases with housing density, also due to the availability of anthropogenic food resulting from littering, waste and deliberate feeding (Lim and Sodhi, 2004). Conversely, insectivores tend to be disadvantaged in urban areas because the occurrence and abundance of their invertebrate prey are dependent on vegetation cover, type and volume, which vary hugely across a city and can be very scarce in many habitats (Clergeau et al., 1998). Among nesting guilds, birds nesting in treetops tend to be less affected by the disturbance associated with urbanisation than are ground or shrub nesters (Fernández-Juricic, 2000).

In tropical Singapore, frugivores rather than omnivores were favoured in areas of low-density housing because of the prevalence of fruit-bearing ornamental plants on housing estates, whereas insectivores and carnivores were adversely affected by increased urbanisation (Lim and Sodhi, 2004). Usually omnivores thrive in urban landscapes, but in Singapore, this guild contained many native species that did not exploit anthropogenic food, so there was no increase in species richness with increasing urbanisation.

Dispersal capability

There are many barriers to dispersal in a built environment, such as busy roads and large, sparsely vegetated areas subject to human disturbance. Species vary in their willingness and propensity to disperse across gaps and hostile habitat; flightless invertebrates can have very limited distributions, and even some small flighted birds are very reluctant to cross gaps. Reluctance to cross habitat gaps is the main reason why some British butterflies (Thomas, 2000; Warren et al., 2001) are prone to extinction in modified landscapes. For those species that are limited to patches of natural vegetation, unable to disperse to other patches, the quality of the vegetation and the configuration of the patches ultimately decide their fate. If the habitat quality is poor, reproduction is also poor, in which case the population can be maintained only through immigration from other patches where conditions are better. However, if the patch is too isolated from such source populations so that no immigration or emigration occurs, that population will die out. If the habitat is good, and predators are not over-abundant, successful breeding is possible, and the population will sustain itself, producing new individuals that will disperse to other patches. In successful urban species, a meta-population structure will often exist, where dispersal between sub-populations existing within patches of varying quality ensures the continuing presence of the species in a city.

A prime example of an urban-adapted species is the house sparrow, which is synanthropic to the extent that densities are positively correlated with the density of human settlement (van Heezik and Adams, 2016). House sparrows are often among the most common species found in urban areas (e.g., van Heezik and Adams, 2016). They fit the 'winner' profile by being granivorous and by nesting in human dwellings. However, even urbanophilic species, such as house sparrows, have suffered recent population declines over the last few decades in the UK and Europe in both urban and rural landscapes (Summers-Smith, 2003). The sparrow's sedentary nature and inability to adapt its home range behaviour may have prevented individuals from adapting to rapid changes in the distribution of critical resources in urban areas, leading to reduced nutritional condition (Vangestel et al., 2010). Urban house sparrows occupy smaller home ranges than their rural conspecifics, particularly in areas where cover is patchy. They live in smaller flocks that have to move longer distances between patches of suitable habitat, exposing them to greater predation risk. Body condition of urban house sparrows is often poorer when hedges and dense bushes become more sparsely distributed. Suburban and rural house sparrows have better body condition than sparrows living in other urban habitats.

Tameness Animals encountered in urban areas often seem to be relatively tame. It is frequently possible to get quite close to them, and they continue to forage or interact with each other despite our presence. Such apparent nonchalance could arise because individuals have become habituated to people through repeated exposure to them, but there are also innate differences in boldness between species (Sol et al., 2002) and consistent differences between individuals of the same species, which reflect individual variation in temperament or personality (Evans et al., 2010; Reale et al., 2007). A high level of boldness, which is associated with high exploration, innovation, aggression, risk-taking and low neophobia (wariness of new objects) can predispose a species to successfully exploiting urban environments (Møller, 2010; Evans et al., 2010; Vines and Lill, 2015) (Figure 3.6

Figure 3.6 Some species actively exploit human food sources: a kookaburra checking out the food at a café in Australia.

Figure 3.7 Though not an urban species, the now endangered quokkas resident on Rottnest Island, Australia have no fears about entering buildings to seek food. This boldness can not only favour the adapter species but also acts to encourage children's curiosity in interacting with wild species.

and 3.7). Studies investigating boldness often measure the flight initiation distance (FID), i.e., the distance from an approaching person at which the species observed initiates escape. Urban individuals tend to flee at shorter approach distances and are less likely to fly than are their exurban conspecifics (Møller, 2009; Vines and Lill, 2015). In song sparrows (*Melospiza melodia*), males showed consistent individual differences in boldness, measured as FID (Evans et al., 2010). Responses from bolder individuals to risky situations are different and can affect the fitness of that individual.

Challenges and responses to urban living

Wild animals do not only cope with urban living but can show adaptations in their behaviour or even in their physiological responses to urban environmental conditions, ranging from different climatic conditions to the presence of artificial structures. Heat island effects, artificial watering throughout dry periods, anthropogenic sources of food (feeding of wildlife by humans, landfill waste) and plantings of a wide variety of non-native species such as ornamental fruiting shrubs mean longer growing seasons and greater food availability. More food allows some species to persist at higher densities but can also result in reduced migratory behaviour, a prolonged breeding season, greater longevity, prolonged circadian activity, changes in nesting and feeding habits and tameness towards people (Luniak, 2004). Birds might be less inclined to migrate from cities that do not experience the same cold conditions and reduced food availability that normally are a feature of temperate winters.

Some species in urban areas start breeding earlier and breed for longer periods than their rural counterparts; examples in Central Europe are urban blackbirds (*Turdus merula*), which breed 1–4 weeks earlier than rural ones, and mallards (*Anas platyrhynchus*) (Luniak, 2004). Favourable food and climatic conditions are also associated with birds living longer, especially if predatory raptors are absent. Urban blackbirds live 1–1.8 years longer than do rural birds (Luniak, 2004). Home ranges of some species may be reduced in size in urban areas if resources are abundant. For example, the small home ranges of badgers (*Meles meles*) radio-tracked in Brighton in the UK and of racoons in the US were attributed to the rich food resources provided by gardens (Davison et al., 2009). Other animals, such as coyotes, expand their home ranges, become less active during hours of human activity and avoid developed areas during the daytime (Tigas et al., 2002).

Extended growth seasons do not always lead to greater productivity. The breeding success of urban house sparrows is lower than that of their rural conspecifics, possibly because urban birds are exposed to higher predation, pollutants from traffic and a paucity of invertebrate prey due to vegetation removal. Chick survival too tends to be lower in cities (Peach et al., 2008). In Hungary, urban house sparrows are smaller than their rural counterparts, and this difference persisted when birds were kept in aviaries in identical conditions and fed ad lib food (Liker et al., 2008). This outcome suggests there could be habitat-related differences in nestling development or adaptive divergence due to distinct environmental conditions, such as differences in predation pressure, driving this gradient in body condition.

Those species showing the greatest adaptability are the ones most likely to thrive in cities. Species that are behaviourally flexible generally are more successful in novel urban habitats (Sol et al., 2002), probably because they are able to find new ways to exploit resources such as food sources and nest sites that are highly variable in space and over time. Successful avian invaders tend to have relatively larger brains and demonstrate innovative foraging in their region of origin (Sol et al., 2002). Coping with human activities, which peak at set times during the day, is achieved by some animals, which alter the times and places at which they forage. For example, European hedgehogs (*Erinaceus europaeus*) change their nocturnal foraging so that they are more active after midnight when there are fewer people and vehicles about and avoid foraging near roads, although they didn't avoid crossing roads (Dowding et al., 2010).

Buildings

Cities by definition are composed of buildings. Collisions with man-made structures are ranked among the top threats to North American birds in terms of numbers killed (Figure 3.8). Erickson et al.'s (2005) summary of collisions in the US reported mortality associated with some tall buildings. For example, over 4,707 dead birds were recovered from around several buildings in Florida over a 3-day period in October 1964. In the autumn of 1970, 707 dead birds were documented below the Empire State Building in New York, and large numbers of nocturnal migrating birds were recorded at the Washington Monument in Washington, DC. Records from long-term studies include 5,454 dead birds at 54 tall glass buildings over a three-year period, with an estimated 733 birds (mostly passerines) killed per building per year. Standardized searches around 5,500 residential houses during the 1989–1990 winter in the US revealed 995 dead birds of 66 species, mostly passerines, an average of 0.85 birds killed per house each winter. These results could be biased upwards because most of the building and window collision data come from studies of known or suspected problem structures, and studies using random selection of buildings are necessary to gain a better understanding of the extent of the mortality.

Figure 3.8 This was one of a number of dead parrots killed presumably from collisions with the overhead rail line. Children will inevitably encounter dead wildlife resulting from such collisions and as roadkill.

Buildings and man-made structures can have positive impacts for some species, by providing opportunities, and they have even allowed vulnerable species to inhabit urban environments. Little penguins (*Eudyptula minor*) nest under buildings and in rocky crevices along artificial breakwaters (Bourne and Klomp, 2004), and brushtail possums (*Trichosurus vulpecula*) and chipmunks (*Tamias* spp.) frequently use buildings for denning (Stratham and Stratham, 1997; Ryan and Larson, 1976). House sparrows use buildings for nest sites to the extent that improvements in building materials resulting in fewer potential nest sites has been proposed as one of the causes behind the decline in house sparrow numbers in more affluent suburbs (Shaw et al., 2008).

Roads

Roads are found virtually everywhere and can be a major constraint to movement of both children and wildlife alike. In Europe, almost a quarter of all land area is located within 500 m of a road or railway line, giving wildlife fewer options to occupy remote areas and contributing to population declines in some species (Torres et al., 2016). Roads are particularly concentrated in urban areas, where they are a dominant feature across the urban landscape, designed to transport people and goods as efficiently as possible, but resulting in ecological impacts such as loss and fragmentation of habitat, input of pollutants, and the creation of barriers to wildlife movements. In Singapore, increasing density of roads reduced the species richness of birds and butterflies, with road avoidance likely explaining the lower abundance of mobile species, such as birds and mammals, with the creation of eddy currents by passing cars possibly affecting butterflies (Chong et al., 2014). Roads cause direct mortality of wildlife due to collisions with vehicles. For example, in Australia, nine out of 23 tagged brush-tailed possums were killed or injured through vehicle collisions during 444 days of radio-tracking (Stratham and Stratham, 1997). In the case of bats, low-flying species are more vulnerable to collisions with vehicles, but also vulnerable are the males and juveniles of any bat species, especially where roads were close to or crossed other linear features, such as rivers, treelines and viaducts (Fensome and Mathews, 2016). Vulnerability to collisions varies between bat species and even between age and sex in the same species. Population-level impacts could occur if dispersal between patches was reduced by collision mortality or specific groups of animals were particularly vulnerable, such as reproductively active females.

Roads also can act as barriers or filters to movement. Tremblay and St. Clair (2009) tested the willingness of forest birds to cross roads, railways, transport bridges across riparian corridors and rivers in an urban area by playing an avian mobbing call as a lure and found that the size of the vegetation gap was the most important predictor of movement in black-capped chickadees (*Poecile atricapillus*), with traffic volume also reducing likelihood of movement. The threshold size of the gap was 45 m, implying that planting trees along the verges and median strips of gaps wider than this could promote connectivity. Variation between species in the degree to which they are affected by roads depends on their ability to avoid cars, the size of their movement ranges and whether reproductive rates and densities were naturally low. Those that respond positively to roads use roads as a food resource and are able to avoid traffic. Roads themselves can also provide habitat, in the form of verges, buffers for aesthetic purposes to screen out commercial land uses or street trees.

Artificial illumination

Cities are bright places. Ecological light pollution includes direct glare, chronically increased illumination and temporary, unexpected fluctuations in lighting from sources such as sky glow, lighted buildings and towers, streetlights, security lights and vehicle lights (Longcore and Rich, 2004). Chronic lighting sources are street lighting, lighting around public places and commercial lighting (advertising), whereas fluctuations occur when sports grounds are floodlit for events. Residents themselves contribute by illuminating their gardens for security or prestige. These all contribute to a high level of light in the urban environment.

Impacts on animals occur because of changes in orientation and attraction or repulsion, which will affect foraging, reproduction, migration and communication (Longcore and Rich, 2004). Some species can extend their foraging period into the night by foraging under artificial lights, and while this seems beneficial for those species, it may not be so for their prey. Animals adapted to navigating in the darkness can be disoriented by constant lighting. Artificial lights can attract and cause mortality directly through collisions or indirectly. Because birds are reluctant to fly out of the lighted area into the dark, they become trapped and flap around the beam of light until exhausted, at which point they are vulnerable to predators, which may choose to hunt near lights if they learn to associate these places with easy prey. Many flying insects are also attracted to light, although this depends on the spectrum of light: some might fly directly onto a hot glass lamp cover and die, whereas others fly around the light until exhausted, flutter to the ground and die or become prey. Responses vary, with some species being repelled by light.

Reproduction can be affected by artificial light. For example, female frogs (*Physalaemus pustulosus*) are less selective when choosing a mate when light levels are increased, presumably preferring to mate quickly and avoid the increased predation risk of mating activity (Rand et al. 1997). Although less studied, artificial lighting is sufficiently bright to affect the timing of leaf burst, flowering and fruiting in plants, growth form and resource allocation and, consequently the physiology, behaviour and ecology of herbivores and pollinators (Bennie et al., 2016). In urban bats (*Pipistrellus pipistrellus*), which are distributed across many European cities, street lighting creates barriers to the ability of this species to move through the city, since bats are less likely to cross gaps when they are well lit (Hale et al., 2015). Networks of trees support bat movements but only if the gaps between them are less than 80 m; however, access to this habitat is further constrained by street lighting. At the community level, by interfering with competition and predator-prey relationships, artificial lighting changes community composition and disrupts ecological functions.

Different types of lighting affect animals differently because they emit radiation in different spectra: older-style sodium lamps are less attractive to insects than are mercury lamps, which emit radiation in both the ultraviolet and blue-green spectra and are particularly attractive to insects. Moths in particular are well-known for their attraction to lights. In Birmingham, UK, Plummer et al. (2016) showed that moths are attracted to suburban gardens with denser and more spectrally diverse street lighting and that a switch from narrow (low-pressure sodium) to broad (high-pressure sodium) spectrum lighting increased the diversity of the larger species of moths in gardens. The concern is that if gardens do not have the resources necessary for moth survival and reproduction, they may act as ecological traps, with the superficial attractiveness of the vegetation actually drawing animals into unsuitable areas.

Artificial lighting in many cities is moving towards light-emitting diode (LED) lighting, which is more energy-efficient and distributes light more evenly. However, the wide range of wavelengths transmitted by LEDs, particularly in the blue part of the spectrum, will overlap with the spectral sensitivities of many visual and biological processes, will exacerbate sky glow and can penetrate deeper into water columns, thereby influencing the circadian rhythms of many species and acting as a powerful attractant (Stanley et al., 2016).

Anthropogenic noise

Cities are noisy, and communicating in this environment is another challenge faced by many species. Anthropogenic noise, such as road traffic, differs from natural noise in that it has an energy spectrum that is mostly concentrated at low frequencies, which reduces the distance at which individuals can properly communicate with each other. Foraging efficiency can be reduced, either directly by reducing the ability of the predator to pick up on auditory cues from their prey or because more time is spent in vigilance and anti-predator behaviours if noise is perceived as a threat, resulting in missed opportunities. For example, continuous noise in the environment of the chaffinch (*Fringilla coelebs*) impairs auditory surveillance, which triggers increased visual surveillance, resulting in less time spent foraging (Quinn et al., 2006). Mate attraction and territorial behaviour can also be affected (Francis and Barber, 2013).

Most of the research on impacts of noise in urban areas has been done on birds because their vocalisations are essential for reproduction and survival. Birds call to advertise themselves to potential mates but also to warn of the approach of a predator. Birds can adapt to a noisy environment by avoiding living in particularly noisy places; by singing at different times of the day; by changing their song in the short term, such as singing louder or at higher frequencies; or by making long-term adaptive changes to their songs (Gross et al., 2010; Slabbekoorn and den Boer-Visser, 2006). Evidence shows that animals abandon areas when frequent or chronic noise interferes with their ability to pick up auditory cues or when sounds are perceived as threats (reviewed in Francis and Barber, 2013). These cues might be calls from conspecifics, but they can also be sounds made by their prey. For example, greater mouse-eared bats (*Myotis myotis*) locate terrestrial prey based on the sounds made when prey are walking and avoid hunting in noisy areas (Schaub et al., 2008). Spatial avoidance by some species less tolerant to noise could result in different animal communities in noisy parts of town. Temporal noise avoidance, or singing at times of the day when there is less background noise, results in a phenomenon called 'gap calling' behaviour, when animals such as frogs call during breaks in anthropogenic noise (Potvin, 2017). In some cases, there are complete shifts in the timing of choruses to avoid peaks in noise from traffic (Fuller et al., 2007b, but see Yang et al., 2014). European robins (*Erithacus rubecula*) have shifted their song to night-time to avoid acoustic interference from daytime urban noise; the consequences of this shift are unknown (Fuller et al., 2007b).

When it is not feasible to avoid calling during noisy periods, some animals have been shown to adjust the amplitude, frequency and duration of calls to avoid being drowned out by background noise (Potvin, 2017). For example, many urban birds vocalise at higher frequencies than their rural counterparts (reviewed in Slabbekoorn, 2013). While such changes permit individuals to communicate in noisy conditions (e.g., Fuller et al., 2007b; Gross et al., 2010), these shifts could also incur a cost, in that males might end up calling at frequencies that are less preferred by females. Female great tits (*Parus major*) can more easily detect male songs sung at higher frequencies in noisy areas, but these males experience higher rates of cuckoldry (Halfwerk et al., 2011a). Great tits breeding in noisy areas have smaller clutches and fewer fledglings (Halfwerk et al., 2011b). Species that sing at lower frequencies might avoid settling in urban environments or have lower breeding success than species with high singing frequencies. In this way, urban noise could act as a filter, determining which species can colonise and persist in urban environments (Cardoso, 2014).

Conclusions and future challenges

It is clear that the kind of nature that children are likely to encounter regularly in urban areas is going to be highly modified, since urbanisation acts as a filter. The more highly urbanised the area, the lower the species diversity and habitat quality, until you are left with only those species that aren't too bothered by human presence and that can find nesting sites and enough food when the vegetation is sparse and not very diverse. Even the larger green spaces, such as parks, tend to be substantially modified. The species that thrive in the most built-up parts of the cities are those that can exploit anthropogenic food sources and are prepared to build their nests on, in or near buildings. What is not clear is whether the type of nature experience influences the quality of the well-being benefit to the child. Associations between green spaces and children's health have sometimes been based on fairly simplistic representations of nature, such as the amount of green cover as detected by remote sensing. Among adults, there is some evidence that suggests that perceptions of psychological well-being benefits do increase with increasing species richness of plants, habitats and numbers of birds, and this may be the case for children, but that remains to be investigated. If this was the case, motivation to improve children's well-being through exposure to more ecologically authentic nature could also result in really positive biodiversity outcomes. However, despite the modified character of nature in cities, nature is still available to children in the small spaces, particularly backyards, where close encounters with plants, birds and invertebrates can occur and natural processes can be observed.

Future challenges for urban nature will inevitably reveal themselves. Stanley et al. (2016) used a horizon-scanning exercise to identify emerging threats to urban ecosystems associated with developing technologies and increasing societal demands on urban nature. While all of them anticipate loss of biodiversity through ongoing loss of habitat, pollution or disturbances to natural behaviour through diverse advances such as changes in lighting, increased use of solar panels and concrete that has the ability to self-seal its own cracks and crevices, some relate more closely to the future nature experiences of children. For example, as the health benefits of nature contact are realised and there is increasing use of green prescriptions and with growing human populations, the number of visits to available green spaces is likely to intensify. Moreover, as people use green spaces more, there may be pressure to redesign them so they are more appropriate for recreational activities, including making pathways, lawns and built structures; these would reduce the 'wildness' of the space, to the detriment of the wildlife populations currently occupying these spaces.

Note

1 A group of *species* exploiting the same resources or different resources in related ways.

References

Aronson, M.F.J., Handel, S.N., La Puma, I.P., and Clemants, S.E. 2015. Urbanization pronates non-woody species and diverse plant assemblages in the New York metropolitan region. *Urban Ecosystems* 18: 31–45. DOI: 10.1007/s11252-014-0382-z

Barratt, B.I.P., Dickinson, K.J.M., Freeman, C., Porter, S., Johnstone, P.D., Wing, J., and van Heezik, Y. 2015. Biodiversity of Coleoptera and other invertebrates in urban gardens: A case study in a New Zealand city. *Insect Conservation and Diversity* 8(5): 428–437. DOI: 10.1111/icad.12120

Barrett, D. Recio, M.R., Barratt, B.I.P., Seddon, P.J., and van Heezik, Y. 2016. Resource selection by an ancient taxon (Onychophora) in a modern urban landscape: A multi-scale analysis approach to assist in the conservation of an animal phylum. *Landscape and Urban Planning* 148: 27–36.

Bennie, J., Davies, T.W., Cruse, D., and Gaston, K.J. 2016. Ecological effects of artificial light at night on wild plants. *Journal of Ecology* 104(3): 611–620.

Blair, R. 2004. The effects of urban sprawl on birds at multiple levels of biological organization. *Ecology and Society* 9(2). www.ecology and www.society.org/vol9/iss5/art2/

Botkin, D.B., and Beveridge, C.E. 1997. Cities as environments. *Urban Ecosystems* 1: 3–19.

Bourne, J., and Klomp, N.I. 2004. Ecology and management of the little penguin *Eudyptula minor* in Sydney Harbour. In D. Lunney and S. Burgin (Eds.), *Urban Wildlife: More Than Meets the Eye*, Royal Zoological Society of New South Wales, Mosman, pp. 131–137.

Burghardt, K.T., Tallamy, D.W., and Shriver, W.G. 2008. Impact of native plants on bird and butterfly biodiversity in suburban landscapes. *Conservation Biology* 23(1): 219–224. DOI: 10.1111/j.1523-1739.2008.01076.x

Cardoso, G.C. 2014. Nesting and acoustic ecology, but not phylogeny, influence passerine urban tolerance. *Global Change Biology* 20: 803–810.

Chong, K.Y., Teo, S., Kurukulasuriya, B., Chung, Y.F., Rajathurai, S., and Tan, H.T.W. 2014. Not all green is as good: Different effects of the natural and cultivated components of urban vegetation on bird and butterfly diversity. *Biological Conservation* 171: 299–309.

Clergeau, P., Jokimaki, J., and Snep, R. 2006. Using hierarchical levels for urban ecology. *Trends in Ecology and Evolution* 21(12): 660–661.

Clergeau, P., Savard, J.-P., Mennechez, G., and Falardeau, G. 1998. Bird abundance and diversity along an urban-rural gradient: A comparative study between two cities on to different continents. *The Condor* 100(3): 413–425.

Colding, J., Lundberg, J., and Folke, C. 2006. Incorporating green area user groups in urban ecosystem management. *Ambio* 35(5): 237–244.

Davison, J., Huck, M., Delahay, R.J., and Roper, T.J. 2009. Restricted ranging behaviour in a high-density population of urban badgers. *Journal of Zoology* 277: 45–53.

Derraik, J.G.B., Rufaut, C.G., Closs, G.P., and Dickinson, K.J.M. 2005. Ground invertebrate fauna associated with native shrubs and exotic pasture in a modified rural landscape, Otago, New Zealand. *New Zealand Journal of Ecology* 29: 129–135.

Dowding, C.V., Harris, S., Poulton, S., and Baker, P.J. 2010. Nocturnal ranging behaviour of urban hedgehogs, *Erinaceus europaeus*, in relation to risk and reward. *Animal Behaviour* 80: 13–21.

Erickson, W.P., Johnson, G.D., and Young, D.P., Jr. 2005. A Summary and Comparison of Bird Mortality from Anthropogenic Causes with an Emphasis on Collisions. USDA Forest Service Gen. Tech. Rep. PSW-GTR-191.

Evans, J., Boudreau, K., and Hyman, J. 2010. Behavioural syndromes in urban and rural populations of song Sparrows. *Ethology* 116(7): 588–595. DOI: 10.1111/j.1439-0310.2010.01771.x

Faeth, S.H., Bang, C., and Saari, S. 2011. Urban biodiversity: Patterns and mechanisms. *Annals of the New York Academy of Sciences* 1223: 69–81.

Fensome, A.G., and Mathews, F. 2016. Roads and bats: A meta-analysis and review of the evidence on vehicle collisions and barrier effects. *Mammal Review* 46: 311–323.

Fernández-Juricic, E. 2000. Avifaunal use of wooded streets in an urban landscape. *Conservation* Biology 14: 513–521.

Francis, C.D., and Barber, J.R. 2013. A framework for understanding noise impacts on wildlife: An urgent conservation priority. *Frontiers in Ecology and the Environment* 11(6): 305–313.

French, K., Major, R., and Hely, K. 2005. Use of native and exotic garden plants by suburban nectarivorous birds. *Biological Conservation* 121: 545–559. http://dx.doi.org/10.1016/j.biocon.2004.06.004

Fuller, R.A., Irvine, K.N., Devine-Wright, P., Warren, P.H., and Gaston, K.J. 2007a. Psychological benefits of greenspace increase with biodiversity. *Biology Letters* 3(4): 390–394.
Fuller, R.A., Tratalos, J., and Gaston, K.J. 2009. How many birds are there in a city of about a million people? *Diversity and Distributions* 15: 328–337.
Fuller, R.A., Warren, P.H., and Gaston, K.J. 2007b. Daytime noise predicts nocturnal singing in urban robins. *Biology Letters* 3(4): 368–370. DOI: 10.1098/rsbl.2007.0134
Gagné, S.A., and Fahrig, L. 2011. Do birds and beetles show similar responses to urbanization? *Ecological Applications* 21: 2297–2312.
Goddard, M.A., Dougill, A.J., and Benton, T.G. 2010. Scaling up from gardens: Biodiversity conservation in urban environments. *Trends in Ecology & Evolution* 25(2): 90–98.
Grimm, N.B. 2008. Global change and the ecology of cities. *Science* 319: 756. DOI: 10.1126/science.1150195
Gross, K., Pasinelli, G., and Kunc, H.P. 2010. Behavioral Plasticity allows short term adjustment to a novel environment. *The American Naturalist* 176(4): 456–464.
Hale, J.D., Fairbrass, A.J., Matthews, T.J., Davies, D., and Sadler, J.P. 2015. The ecological impact of city lighting scenarios: Exploring gap crossing thresholds for urban bats. *Global Change Biology* 21: 2467–2478.
Halfwerk, W., Bot, S., Buikx, J., van der Velde, M., Komdeur, J., ten Cate, C., and Slabbekoorn, H. 2011a. Low-frequency songs lose their potency in noisy urban conditions. *Proceedings of the National Academy of Science USA* 108: 14549–14554.
Halfwerk, W., Holleman, L.J.M., Lessells, C.M., and Slabbekoorn, H. 2011b. Negative impact of traffic noise on avian reproductive success. *Journal of Applied Ecology* 48: 210–219.
Helden, A.J., Stamp, G.C., and Leather, S.R. 2012. Urban biodiversity: Comparison of insect assemblages on native and non-native trees. *Urban Ecosystems* 15(3): 611–624.
Hodgkison, S.C., Hero, J.-M., and Warnken, J. 2006. The conservation value of suburban golf courses in a rapidly urbanising region of Australia. *Landscape and Urban Planning* 79(3–4): 323–337.
Jokimäki, J., and Suhonen, J. 1998. Distribution and habitat selection of wintering birds in urban environments. *Landscape and Urban Planning* 39(4): 253–263.
Jones, E.L., and Leather, S.L. 2012. Invertebrates in urban areas: A review. *European Journal of Entomology* 109: 463–478.
Kark, S., Iwaniuk, A., Schalimtzek, A., and Banker, E. 2007. Living in the city: Can anyone become an 'urban exploiter'? *Journal of Biogeography* 34: 638–651.
Kelly, D., Ladley, J.J., Robertson, A.W., Anderson, S.H., Wotton, D.M., and Wiser, S.K. 2010. Mutualisms with the wreckage of an avifauna: The status of bird pollination and fruit-dispersal in New Zealand. *New Zealand Journal of Ecology* 34(1): 66–85.
Knapp, S., Dinsmore, L., Fissore, C., Hobbie, S.E., Jakobsdottir, I., Kattge, J., King, J.Y., Klotz, S., McFadden, J.P., and Cavender-Bares, J. 2012. Phylogenetic and functional characteristics of household yard floras and their changes alien an urbanization gradient. *Ecology* 93(8): S83–S98. http://dx.doi.org/10.1890/11-0392.1
Kowarik, I., Buchholz, S., von der Lippe, M., and Seitz, B. 2016. Biodiversity functions of urban cemeteries: Evidence from one of the largest Jewish cemeteries in Europe. *Urban Forestry & Urban Greening* 19: 68–78.
Liker, A., Papp, Z., Bókony, V., and Lendvai, Á.Z. 2008. Lean birds in the city: Body size and condition of house sparrows along the urbanization gradient. *Journal of Animal Ecology* 77(4): 789–795.
Lim, H.C., and Sodhi, N.S. 2004. Responses of avian guilds to urbanisation in a tropical city. *Landscape and Urban Planning* 66: 199–215.

Longcore, T., and Rich, C. 2004. Ecological light pollution. *Frontiers in Ecology and the Environment* 2(4): 191–198.

Loram, A., Thomspon, K., Warren, P.H., and Gaston, K.J. 2008. Urban domestic gardens (XII): The richness and composition of the flora in five UK cities. *Journal of Vegetation Science* 19(3): 321–330.

Luniak, M. 2004. Synurbization: Adaptation of Animal Wildlife to Urban Development. Proceedings of the 4th International Urban Wildlife Symposium, Shaw et al., Eds. Pp. 50–54.

Madre, F., Vergnes, A., Machon, N., and Clergeau, P. 2013. A comparison of 3 types of green roof as habitats for arthropods. *Ecological Engineering* 57: 109–117.

Magle, S.B., Lehrer, E.W., and Fidino, M. 2015. Urban mesopredator distribution: Examining the relative effects of landscape and socioeconomic factors. *Animal Conservation* 19(2): 163–175.

Matteson, K.C., Ascher, J.S., and Langellotto, G.A. 2008. Bee richness and abundance in New York City urban gardens. *Annals of the Entomological Society of America* 101(1): 140–150.

Matteson, K.C., and Langellotto, G.A. 2010. Determinates of inner city butterfly and bee species richness. *Urban Ecosystems* 13(3): 333–347.

McDonnell, M.J., and Hahs, A.K. 2008. The use of gradient analysis studies in advancing our understanding of the ecology of urbanizing landscapes: Current status and future directions. *Landscape Ecology* 23(10): 1143–1155.

McIntyre, N.E., and Rango, J.J. 2009. Arthropods in urban ecosystems: Community patterns s functions of anthropogenic land use. In M.J. McDonnell, A.K. Hahs, and J.H. Breuste (Eds.), *Biology of Cities and Towns: A Comparative Approach*, Cambridge University Press, Cambridge, pp. 233–242.

McKinney, M.L. 2008. Effects of urbanization on species richness: A review of plants and animals. *Urban Ecosystems* 11(2): 161–176.

McKinney, M.L., and Lockwood, J.L. 1999. Biotic homogenization: A few winners replacing many losers in the next mass extinction. *Trends in Ecology and Evolution* 14(1): 450–453. http://dx.doi.org/10.1016/S0169-5347(99)01679-1

Møller, A.P. 2009. Successful city dwellers: A comparative study of the ecological characteristics of urban birds in the Western Palearctic. *Oecologia* 159: 849–858.

Møller, A.P. 2010. Interspecific variation in fear responses predicts urbanization in birds. *Behavioral Ecology* 21(2): 365–371. DOI: 10.1093/beheco/arp199

Niemelä, J., Kotze, D.J., and Yli-Pelkonen, V. 2009. Comparative urban ecology: Challenges and possibilities. In M.J. McDonnell, A.K. Hahs, and J.H. Breuste (Eds.), *Biology of Cities and Towns: A Comparative Approach*, Cambridge University Press, Cambridge, pp. 9–23.

Pardee, G.L., and Philpott, S.M. 2014. Native plants are the bee's knees: Local and landscape predictors of bee richness and abundance in backyard gardens. *Urban Ecosystems* 17: 641–659.

Peach, W.J., Vincent, K.E., Fowler, J.A., and Grice, P.V. 2008. Reproductive success of house sparrows along an urban gradient. *Animal Conservation* 11(6): 493–503.

Plummer, K.E., Hale, J.D., O'Callaghan, M.J., Sadler, J.P., and Siriwardena, G.M. 2016. Investigating the impact of street lighting changes on garden moth communities. *Journal of Urban Ecology* 2(1): 1–10.

Potvin, D. 2017. Coping with a changing soundscape: Avoidance, adjustments and adaptations. *Animal Cognition* 20(1): 9–18. DOI: 10.1007/s10071-016-0999-9

Quinn, J.L., Whittingham, M.J., Butler, S.J., and Cresswell, W. 2006. Noise, predation risk compensation and vigilance in the chaffinch *Fringilla coelebs*. *Journal of Avian Biology* 37: 601–608.

Rand, A.S., Bridarolli, M.E., Dries, L., and Ryan, M.J. 1997. Light levels influence female choice in Tungara frogs: Predation risk assessment? *Copeia* 1997: 447–450.

Reale, D., Reader, S.M., Sol, D., McDougal, I.P.T., and Dingemanse, N.J. 2007. Integrating animal temperament within ecology and evolution. *Biological Reviews* 82: 291–318.

Ryan, D.A., and Larson, J.S. 1976. Chipmunks in residential environments. *Urban Ecology* 2: 173–178.

Schaub, A., Ostwald, J., Björn, M., and Siemers, M. 2008. Foraging bats avoid noise. *Journal of Experimental Biology* 211: 3174–3180. DOI: 10.1242/jeb.022863

Shapiro, A.M. 2002. The Californian urban butterfly fauna is dependent on alien plants. *Diversity and Distributions* 8: 31–40.

Shaw, L.M., Chamberlain, D., and Evans, M. 2008. The house sparrow *Passer domesticus* in urban areas: Reviewing a possible link between post-decline distribution and human socioeconomic status. *Journal of Ornithology* 149: 293–299.

Simao, M.C.M., Flory, S.L., and Rudgers, J.A. 2010. Experimental plant invasion reduced arthropod abundance and richness across multiple trophic levels. *Oikos* 119: 1553–1562.

Slabbekoorn, H. 2013. Songs of the city: Noise-dependent spectral plasticity in the acoustic phenotype of urban birds. *Animal Behaviour* 85(5): 1089–1099.

Slabbekoorn, H., and den Boer-Visser, A. 2006. Cities change the songs of birds. *Current Biology* 16: 2326–2331.

Smith, R.M., Gaston, K.J., Warren, P.H., and Thompson, K. 2006a. Urban domestic gardens (VIII): Environmental correlates of invertebrate abundance. *Biodiversity and Conservation* 15: 2515–2545.

Smith, R.M., Gaston, K.J., Warren, P.H., and Thompson, K. 2006b. Urban domestic gardens (VI): Environmental correlates of invertebrate species richness. *Biodiversity and Conservation* 15: 2415–2438.

Smith, R.M., Thompson, K., Hodgson, J.G., Warren, P.H., and Gaston, K.J. 2006. Urban domestic gardens (IX): Composition and richness of the vascular plant flora, and implications for native biodiversity. *Biological Conservation* 129: 312–322. http://dx.doi.org/10.1016/j.biocon.2005.10.045

Sol, D., Timmermans, S., and Lefebre, L. 2002. Behavioural flexibility and invasion success in birds. *Animal Behaviour* 63: 495–502.

Sorace, A. 2002. High density of bird and pest species in urban habitats and the role of predator abundance. *Ornis Fennica* 79: 60–71.

Stagoll, K., Manning, A.D., Knight, E., Fischer, J., and Lindenmayer, D.B. 2010. Using bird-habitat relationships to inform urban planning. *Landscape and Urban Planning* 98: 13–25.

Stanley, M.C., Beggs, J.R., Bassett, I.E., Burns, B.R., Dirks, K.N., Jones, D.N., Linklater, W.L., Macinnis-Ng, C., Simcock, R., Souter-Brown, G., Trowsdale, S.A., and Gaston, K.J. 2016. Emerging threats in urban ecosystems: A horizon scanning exercise. *Frontiers in Ecology and the Environment* 13(10): 553–560. DOI: 10.1890/150229

Stratham, M., and Stratham, H.L. 1997. Movements and habits of brushtail possums (*Trichosurus vulpecula* Kerr) in an Urban Area. *Wildlife Research* 24(6): 715–726.

Shochat, E., Warren, P.S., Faeth, S.H., McIntyre, N.E., and Hope, D. 2006. From patterns to emerging processes in mechanistic urban ecology. *Trends in Ecology and Evolution* 21: 186–191.

Thomas, C.D. 2000. Dispersal and extinction in fragmented landscapes. *Proceedings of the Royal Society of London Series B: Biological Sciences* 267: 139–145.

Thompson, K., Austin, K.C., Smith, R.M., Warren, P.H., Angold, P.G., and Gaston, K.J. 2003. Urban domestic gardens (I): Putting small-scale plant diversity in context. *Journal of Vegetation Science* 14: 71–78. http://dx.doi.org/10.1111/j.1654-1103

Threlfall, C.G., Williams, N.S.G., Hahs, A.K., and Livesley, S.J. 2016. Approaches to urban vegetation management and the impacts on urban bird and bat assemblages. *Landscape and Urban Planning* 153: 28–39.

Tigas, L.A., Van Vuren, D.H., and Sauvajot, R.M. 2002. Behavioral responses of bobcats and coyotes to habitat fragmentation and corridors in an urban environment. *Biological Conservation* 108(3): 299–306.

Torres, A., Jaeger, J.A.G., and Alonso, J.C. 2016. Assessing large-scale wildlife responses to human infrastructure development. *PNAS* 113(30). DOI: 10.1073/pnas.1522488113

Tóthmérész, B., Máthé, I., Balázs, E., and Magura, T. 2011. Response of carabid beetles to urbanization in Transylvania (Romania). *Landscape and Urban Planning* 101: 330–337.

Tremblay, M.A., and St. Clair, C. 2009. Factors affecting the permeability of transportation and riparian corridors to the movements of songbirds in an urban landscape. *Journal of Applied Ecology* 46(6): 1314–1322.

Summers-Smith, J.D. 2003. The decline of the house sparrow: A review. *British Birds* 96: 439–446.

Vangestel, C., Braeckman, B.P., Matheve, H., and Lens, L. 2010. Constraints on home range behaviour affect nutritional condition in urban house sparrows (*Passer domesticus*). *Biological Journal of the Linnaean Society* 101: 41–50.

van Heezik, Y.M., and Adams, A.L. 2016. Vulnerability of native and exotic urban bird to housing densification and changing gardening and landscaping trends. *Urban Ecosystems*. DOI: 10.1007/s11252-014-0379-7

van Heezik, Y.M., Dickinson, K.J.M., Freeman, C., Porter, S., Wing, J., and Barratt, B.I.P. 2016. To what extent does vegetation composition and structure influence beetle communities and species richness in private gardens in New Zealand? *Landscape and Urban Planning* 151: 79–88.

van Heezik, Y.M., Freeman, C., Porter, S., and Dickinson, K.J.M. 2014. Native and exotic woody vegetation communities in domestic gardens in relation to social and environmental factors. *Ecology and Society* 19(4): 17. http://dx.doi.org/10.5751/ES-06978-190417

van Heezik, Y.M., Smyth, A., and Mathieu, R. 2008. Diversity of native and exotic birds across an urban gradient in a New Zealand city. *Landscape and Urban Planning* 87: 223–232.

van Rensburg, B.J., Peacock, D.S., and Robertson, M.P. 2009. Biotic homogenization and alien bird species along an urban gradient in South Africa. *Landscape and Urban Planning* 92: 233–241.

Vergnes, A., Le Viol, I., and Clergeau, P. 2012. Green corridors in urban landscapes affect the arthropod communities of domestic gardens. *Biological Conservation* 145: 171–178.

Vergnes, A., Pellissier, V., Lemperiere, G., Rollard, C., and Clergeau, P. 2014. Urban densification causes the decline of ground-dwelling arthropods. *Biodiversity and Conservation* 23: 1859–1877.

Vines, A., and Lill, A. 2015. Boldness and urban dwelling in little ravens. *Wildlife Research* 42: 590–597.

Warren, M.S., Hill, J.K., Thomas, J.A., Asher, J., Fox, R., Huntley, B., Roy, D.B., Telfer, M.G., Jeffcoate, S., Harding, P., Jeffcoate, G., Willis, S.G., Greatorex-Davies, J.N.,

Moss, D., and Thomas, C.D. 2001. Rapid responses of British butterflies to opposing forces of climate and habitat change. *Nature* 414: 65–69.

WinklerPrins, A.G.A. 2002. Houselot gardens in Santarém, Pará, Brazil: Linking rural with urban. *Urban Ecosystems* 6(1–2): 43–65.

Yang, X.-J., Ma, X.-R., and Slabbekoorn, H. 2014. Timing vocal behavior: Lack of temporal overlap avoidance to fluctuating noise levels in singing Eurasian wrens. *Behavioural Processes* 108: 131–137.

Zhang, H., and Jim, C.Y. 2014. Species diversity and performance assessment of trees in domestic gardens. *Landscape and Urban Planning* 128: 23–34.

4 Human dimensions of urban ecosystems

Human impacts

The power of collective action and its consequences for nature are never more evident than in urban landscapes, where the actions of many people affect the living spaces and availability of resources for many species. Cities are sites where human interactions are the most diverse, both in their nature and in their effects. Human values and attitudes shape their immediate environments through the preferences and actions of countless individuals but also through planning, policies and strategic visions. In this chapter, we discuss, from the perspective of an ecologist, some of the more prevalent human activities that have city-wide impacts for ecosystems and biodiversity: bird feeding, wildlife gardening and recreation. We also explore aspects of urban landscapes that are often under-valued and taken for granted: ecosystem goods and services. In our discussion of the way that people's preferences and values affect ecosystem services and biodiversity, we focus on landforms that are of particular relevance to children: gardens, trees and birds. We conclude by comparing landscape design strategies and consider their value in the context of children's needs.

Challenges and benefits of human interactions with urban wildlife

The previous chapter described a city landscape that poses many challenges for species and natural processes. It functions as a filter to exclude many species that are unable to find the resources they require, are intolerant of high levels of disturbance and pollution or are vulnerable to common urban predators. Despite these limitations, there is still plenty of scope for people not only to interact with wildlife in cities but to provide resources that enhance populations of some species. In Chapter one, we described a continuum of nature to encapsulate the range of nature experiences that are available to children in cities. While it would be preferable that most children have access to ecologically intact, biodiverse habitats that provide the opportunity for truly natural experiences, this is usually not the case; however, children can still engage with nature in meaningful ways through activities such as bird feeding or helping in the garden. The focus on gardens as nature spaces serves to remind us of the huge influence householder preferences have on landscaping, the species they plant or weed out and the activities they engage in, such as bird feeding or building nest boxes, which affects the biodiversity value of these green spaces. Feeding of animals other than birds is common in some countries; for example, brushtail possums (*Trichosurus vulpecula*) in Australia, red foxes (*Vulpes vulpes*) in Switzerland and raccoons (*Procyon lotor*) in the US; however, the general message is that it is better not to do so. Animals that are regularly fed are likely to become less capable and adaptable foragers and become dependent on humans for food, which puts them at risk if the human feeder stops feeding them. They can also lose their fear of humans, which makes them vulnerable to abuse by other people, and increased densities facilitate the spread of diseases and parasites.

Bird feeding

Attracting birds by providing them with food is probably the most popular and widespread form of human-wildlife interaction across much of the western world (Fuller et al., 2008; Robb et al., 2008), and yet it has been associated with few negative effects. Bird feeding allows children to become familiar with birds, witness bird behaviour and think about their place in the food web. More than 50 million Americans feed birds and other wildlife around their homes (US Fish and Wildlife Service, 2001). Proportions of households providing food range from 66% in the US (Lepczyk et al., 2004), 64% across urban and rural areas in England, 53% within five British cities (Davies et al., 2012) and 75% in Leeds, UK (Goddard et al., 2013), to 44.6% in New Zealand and 38–57% in Australia (Rollinson et al., 2003; Howard and Jones, 2004; Galbraith et al., 2014). In Australia, nearly a half of these provide food on a daily basis (Rollinson et al., 2003; Chamberlain et al., 2005), while in British cities, 29% of householders provide food at least once a week (Davies et al., 2012). Varieties of types of food are provided, such as sugar mixes, seeds, bread, meats and food scraps, for a variety of reasons. Where the winter climate is harsh, such as in the northern hemisphere, some people feed to help birds through periods when food may be limited (Figure 4.1). In general, feeding often takes place all year round to enhance reproductive success and just to have birds in our immediate environments to watch and enjoy.

Given the huge scale of bird feeding across urban areas, it is important to understand the benefits but also potential detrimental effects for wildlife populations. Benefits are experienced by both the feeders and the fed. Most people who feed birds do so because it brings them pleasure (Galbraith et al., 2014). Cox and Gaston (2016) found the people who fed birds in the UK felt relaxed and connected to nature when viewing the birds in their gardens. Some of them are concerned that there is not enough natural food available in an urban environment. No studies have focused on children and what they experience when feeding birds, even though activities such as feeding the ducks are common among children. Regardless of

Figure 4.1 Bird feeding through the provision of sugar water is a popular way of attracting native species into New Zealand gardens.

age, feeding birds provides an important way of engaging people with nature in urban areas on a daily basis, which benefits both people and conservation through supporting bird populations and fostering environmental awareness and guardianship (Figure 4.2).

The consequences of providing food for the bird populations are varied and not all necessarily detrimental, but more research is necessary. Bird feeding can influence the survival of some species as well as the timing of reproduction. Supplemented birds typically nest earlier and are more likely to re-nest (Chamberlain et al., 2005), but they do not necessarily produce more fledglings for reasons that are still not clear (reviewed in Jones, 2011). Some species benefit from feeding more than others – i.e., those bold enough to use a feeder and those that will eat the types of food commonly provided. Aggressive birds at feeders hinder access to food by other species. For example, invasive ring-necked parakeets (*Psittacula krameri*) in Paris, France, prevented the native European starling (*Sturnus vulgaris*) from accessing food (Le Louarn et al., 2016). Most birds that use feeders are already common urban exploiters, so that while the provision of food might enhance the abundance of these species, the number of species in the neighbourhood of the feeders tends to remain unchanged (Fuller et al., 2008). Increases in abundance of some species sometimes cause decreases in others. In Auckland, New Zealand, several exotic species benefited from supplementary feeding, at the expense of one native species, probably because of increased

Figure 4.2 Cages in the market at Den Passar, Indonesia, are filled with birds being sold as household pets. Birds are the most popular pet in Indonesia, where more than a fifth of households have them. Unfortunately, most are trapped illegally, and conservationists are concerned about impacts on protected species and on local and regional populations.

levels of disturbance by the exotic species (Galbraith et al., 2015). Provision of food for red kites (*Milvus milvus*) in gardens in Reading, in the UK, was probably the most important factor explaining their abundance in the city. The red kite has been reintroduced across the UK following near extinction after decades of persecution but was not expected to colonise urban areas (Orros and Fellowes, 2015). However, concerns have been raised about the low nutritional value of the food provided to red kites in urban areas, as it might contain potentially harmful additives and lacks skin and bone, both of which are necessary for healthy growth in young kites.

Some of the consequences of food provision are hard to predict. For example, great tits (*Parus major*) that were provided with supplementary food during the breeding season started dawn singing later than birds with no extra food, postponing their regular dawn chorus, although the reason for this was not clear. By delaying their dawn chorus, these males could potentially reduce their reproductive success, if, for example, females based their extra-pair mating decisions on the dawn song performance of their mates (Saggese et al., 2011). Supplementary feeding of birds might also affect other populations of wildlife. Orros et al. (2014)

showed that beetle populations close to bird feeders were depleted relative to matched areas some distance away, due to higher activity under the feeding stations by ground-foraging birds. Feeders situated within one to five metres of windows can result in increased bird collisions, especially during the migration season (Kummer and Bayne, 2015). Other possible detrimental effects include reliance on an unpredictable resource, reduction in diet quality and loss of natural foraging behaviours (reviewed in Davies et al., 2012). To date, bird feeding has been linked to two disease epidemics: conjunctivitis among American house finches (*Carpodacus mexicanus*) in the eastern US and the protozoan disease *Trichosomoniasis gallinae* in the UK, but evidence for the role of feeders in spreading disease is mixed (Jones, 2011).

Wildlife gardening

Besides vegetation, gardens contain features such as ponds, compost heaps, woodpiles, hedges or fences that provide resources and habitat for wildlife (Davies et al., 2009). The responses we collected from children about nature (Chapter 2) frequently mentioned their own gardens and gardening activities. *Wildlife gardening* or *ecological gardening* involves gardening practices that both minimise negative impacts (minimal use of fertiliser, pesticides and weeding and use of organic compost) and enhance habitat quality through creating a diversity of resources. These might include the provision of nest sites for birds and bees and ponds for amphibians (Gaston et al., 2007; Davies et al., 2009). In New Zealand, backyard biodiversity initiatives encourage householders to trap and kill invasive species that prey on wildlife, such as rats (*Rattus* spp.) and brushtail possums. Children are often aware of the problems that introduced pests pose to native species in New Zealand, as illustrated by the quote from Chapter 2: "*Yes. I have 2 cats and they are amazing, but they sometimes catch birds. There are also a lot of possums around my house. Although my cats kill birds, they also kill other pests like rats, mice, and even a baby possum once.*" Wildlife gardening does seem to have a positive effect on species diversity (Goddard et al., 2013). In the UK, gardens with ponds and those where efforts had been made to attract wildlife contained more species (Davies et al., 2011). In Switzerland, wildlife gardening was also associated with greater species diversity, with ecological gardens deemed aesthetically pleasing as long as their appearance was not chaotic (Lindemann-Matthies and Marty, 2013).

The plant species people have in their gardens also affect the kinds of birds they attract. In Australia, for example, the number of rainbow lorikeets (*Trichoglossus haematodus*) has increased hugely across many cities in response to the large number of nectar-bearing trees and shrubs planted in private gardens (White et al., 2005). Gardens in which people have planted sufficient native plants also support native and exotic birds (Daniels and Kirkpatrick, 2006), even from foraging guilds such as insectivores that are generally disadvantaged by urbanisation (Andersson et al., 2007). Gardens are also important habitat for a number of mammals, reptiles and amphibians, and as is the case for birds, garden size, vegetation structure and the proportion of the area that is vegetated all affect the likelihood that wildlife species will be present and thrive.

Motivations behind wildlife gardening in the UK are varied but include personal well-being and satisfaction and a moral responsibility to nature. A significant barrier to the uptake of wildlife-friendly activities are neighbourhood norms, as wildlife gardens can be seen by neighbours as being messy and unattractive (Goddard et al., 2013), but given the contagion in front-yard landscaping observed in some studies (Hunter and Brown, 2012), social norms might act to facilitate wildlife gardening once some critical threshold of number of wildlife gardens along a street is passed. For example, in Fremantle, a suburb of Perth in Western Australia, where water shortages are severe, there was widespread householder adoption of native gardens as a means of reducing water usage and promoting biodiversity. The drivers of this sort of change in landscaping preference are essentially the same as those previously driving European-style gardening practices – knowledge, functionality and social norms. In the Fremantle case, a pro-environmental worldview promoted the acceptance of gardening practices that embraced native biodiversity, and as the number of native gardens increased, so did social acceptance (Uren et al., 2015). When children grow up in this kind of neighbourhood, it becomes normal to them to support and protect native species.

By participating in wildlife gardening, children can connect with nature as they benefit directly from the increase in wildlife resulting from such practices, but they also become more aware of the nature around them. For example, replacement of an exotic invasive plant, such as a buddleia bush, with native plants will attract more native species into the garden. In this New Zealand garden, the buddleia bush was replaced with a native berry-bearing shrub, *Corokia*, which produces fruit that is eaten by native birds (Figure 4.3). Features in wildlife gardens such as wood piles, ponds and hedges also provide children with more play opportunities than a highly manicured garden, with perhaps less emphasis on 'not making a mess'.

Recreation

City dwellers' desire to spend time in natural spaces can threaten the species that live there, if the level of disturbance is too high. Patches of natural habitats are often popular as recreational areas, which can compromise the value of the habitat for wildlife. In response to public demand, city planners develop trail systems for a variety of human recreational uses. Increased pressure on such spaces as human populations increase, along with the popularity of nature health interventions such as *green prescriptions*, prompted Stanley et al. (2015) to propose that increased recreational use of green spaces was an emerging threat to urban biodiversity. Pedestrians, including young children, interrupt bird foraging, causing the birds to flee and reducing the likelihood that they will use busy sites (Fernández-Jurici and Tellería, 2000). In three cities in Mexico, where high numbers of unleashed dogs chase and kill birds, dog presence was negatively related to bird abundance

Figure 4.3 Helping to remove the exotic and invasive buddleia to make way for native Corokia bushes. Corokia berries are used as a food source for native birds. Despite being classed as a pest plant, buddleia can be a food source for native butterflies.

(MacGregor-Fors and Schondube, 2011). Dog presence can mean that birds spend more time being vigilant, waste valuable energy and lose foraging opportunities if they have to repeatedly fly away from dogs. Dogs can also introduce parasites and disease to wildlife populations. While dogs may play a role in enhancing children's nature connection by providing a reason for them to visit natural spaces – i.e., to walk the dog (Chapter 6) – their frequent presence can reduce the value of the nature experience. Laws requiring that dogs be walked on leashes should reduce some forms of disturbance. Forrest and St. Clair (2006) explored the impacts of dog walking on bird and small mammal communities in urban parks in Edmonton, Canada. In contrast to other studies showing that wildlife is affected by free-running dogs (Lafferty, 2001), they found no effects and suggested that the birds in the areas monitored were already tolerant to moderate levels of human activity (Figure 4.4).

Figure 4.4 This bowerbird in Sydney, Australia, has built its bower next to a playground, is highly tolerant of noise and children's activity and benefits from people leaving blue items (commonly straws, pegs and milk bottle tops) for him to add to his bower.

Valuing the urban

Urban ecosystem services

There is growing recognition that urban areas are vital in the provision of ecosystem goods and services, particularly when they are situated in areas that are rich in resources (reviewed in Gaston et al., 2013). Ecosystem services are grouped as follows: provisioning (e.g., urban food production, water), regulating (e.g., local climate, pollution reduction, flood regulation), habitat or supporting (e.g., soil formation and nutrient cycling) and cultural (e.g., aesthetic, recreation, well-being; Millennium Ecosystem Assessment, 2005). Urban ecosystem services provide between US$3,212 and US$17,772 of benefits per hectare per year (Elmqvist et al.,

2015). In comparison, the cost of restoring a degraded urban ecosystem (including planning, preparation, modest soil restoration, plant propagation and planting) was US$26,000 for meadows and $49,000 for woodland. These figures show that it is much more cost-effective to protect and enhance existing green spaces and their associated ecosystem services than to let them degrade to the point where they need to be restored (Elmqvist et al., 2015). Ecosystem services are dependent on many aspects of biodiversity, such as species diversity and composition, population densities, habitat quantity and quality, species movements between habitats and the management of green spaces, as well as activities such as how green waste is managed, use of green walls and roofs, bird feeding and wildlife gardening. All of these influence ecosystem goods and services (Gaston et al., 2013).

With a global emphasis on reducing carbon emissions, signatories to the Kyoto Protocol are required to quantify national organic carbon stocks, including those held in urban areas. Evidence suggests that urban areas not only store more carbon than was previously realised, but they have the potential to store even more if vegetation cover could be increased. Allinson et al. (2016) explored whether household carbon emissions could be offset by gardens and found that while carbon emissions could not be offset by growing trees in gardens, large amounts of organic carbon are stored in householders' gardens.

Pollination of plants by animals is essential for reproduction in many plant species and is an example of how an ecosystem service depends on biodiversity. Pollination is referred to as a regulating service and, when functioning properly, is a service for humans that often goes unnoticed. If pollination rates decline because of fewer pollinators, then agriculture, horticulture and local populations of some plant species could be threatened. Both allotments and private gardens can be particularly suitable areas for wild pollinators because of the large diversity of flowering plants they contain and prolonged availability of nectar (Colding et al., 2006). Networks of smaller patches can support mobile species, such as bees, even when these are physically isolated. However, not all suburbs have diverse gardens; in the case of some butterfly species which have the capacity to move between habitat patches, there are too few suitable patches to support their populations (Wood and Pullin, 2002).

Ecosystem services are influenced strongly by human values and preferences. More than in any other landscape, biodiversity in cities and ecosystem services are the product of both top-down and bottom-up human influences (Kinzig et al., 2005). Humans preserve elements of the original biodiversity in nature reserves, but over most of any cityscape, human activities completely deconstruct natural communities and then rebuild them with largely non-native plants to create recreational parks, gardens and landscapes (Figure 4.5). Top-down influences include national and local legislation and regulations, while bottom-up influences are driven by individual preferences linked to socio-economic status, ethnicity, culture and social norms. Plant communities are maintained through irrigation; weeding; application of fertilizers, herbicides and pesticides; and constant removal and replacement of plants. A general preference for manicured or ordered landscapes means little dead wood is allowed to remain, and weeds are removed. Lawns, which are virtual monocultures of exotic grasses, are a very popular landscape

Figure 4.5 This wildlife garden was created on the site of a house that had to be demolished following the 2011 Christchurch earthquake and demonstrates the vegetation richness that can be achieved on a small piece of land in a short time period.

Figure 4.6 The primarily monocultural lawn does have one huge asset in that it provides a robust landscape for children's play as demonstrated in this photo of the lawns surrounding the Diana Princess of Wales memorial fountain in Hyde Park, London.

feature but require large amounts of water and fertiliser (Figure 4.6). The conversion of native vegetation to parks and gardens consisting of lawn and trees with little understorey has had a large impact on the communities of native animals but also on ecosystem functions, such as nitrogen cycling, water flow and carbon balances (Faeth et al., 2011).

Value of gardens

Gardens (private, community, allotment) together comprise a substantial proportion of urban landscapes, and the biodiversity values and ecosystem services they offer to the city are the result of collective management decisions and preferences of many householders. Private gardens cover between 22% and 27% of the total area of a number of large cities in the UK, 16% of the central part of Stockholm and 36% of the urban area of Dunedin, New Zealand (Loram et al., 2007; Mathieu et al., 2007; Colding et al., 2006).

The evidence for environmental benefits from gardens is mixed. Gardens provide important ecosystem services in the form of food production, carbon sequestration, pollination, the reduction of storm-water runoff, air filtration and mitigation of the heat island effect by shading and through evapo-transpiration (summarised in Cameron et al., 2012 and Dewaelheyns et al., 2016). For example, modelling has shown that a 10% increase in green vegetation across cities, including gardens, which contain about 25% of all urban non-woodland trees, could completely mitigate the 4 degree C temperature rise predicted to occur over the next 80 years in Manchester, in the UK (Gill et al., 2007).

Although gardening is promoted as an environmentally friendly activity, the use of mechanised tools, synthetic fertilisers, pesticides and potable water all have negative impacts. Gardens are also an important source of invasive species, with gardening activities promoting invasive species spread, as well as increasing soil sealing and greenhouse gas emissions (Cameron et al., 2012; Dewaelheyns et al., 2016). While woody plants provide carbon sequestration, the weekly mowing of a highly maintained lawn releases 1.5 times more carbon than is sequestered by the lawn itself (Jo and McPherson, 1995). Fertilisers contribute to greenhouse gas emissions and pesticides to water pollution (summarised in Cameron et al., 2012). Water use in the garden can be heavy during dry periods, as much as 56% of domestic water use in Western Australia (Syme et al., 2004), and climate change will increase this demand in temperate regions.

On the positive side, all forms of urban agriculture support diverse communities of bees and butterflies, which are important for food production (Matteson and Langellotto, 2010; Colding et al. 2006), and the vegetation found in gardens, in particular trees, intercepts rainwater temporarily and mitigates storm-water problems and flooding (Xiao and McPherson, 2002). Gardens also contain large numbers of species, even though the species found in gardens tend to be generalist, mobile and adaptable forms. Populations of these same species in some cases are experiencing difficulties in intensively farmed landscapes. Possibly the greatest benefits enjoyed by people, which derive from the proximity of gardens to the home, are in human health and well-being, as discussed in Chapter 7. This is particularly true for children, whose movements are often restricted to their gardens. Some children in our *Natural Neighbourhoods for City Children* study who were not allowed to leave their own gardens unaccompanied still had good knowledge about nature and natural processes if their home gardens were diverse in habitats and species. Children benefit from naturalistic gardens for play, exploration, manipulating natural elements such as soil and creating structures (Figure 4.7).

The decisions made by householders have a direct influence on the ecosystem services and biodiversity values of gardens. These can be negative, such as the

Figure 4.7 Gardens provide excellent opportunities for creative play such as digging.
Photo by Alex Sims

widespread conversion of permeable front gardens to car-parking surfaces, which has resulted in a 22.5% increase in impermeable front gardens in Southampton between 1991 and 2011 (Warhurst et al., 2014) and a 13% increase in Leeds over a 33-year period (Perry and Nawaz, 2008). In Southampton, by 2011, 55% of front gardens were paved over, a little less than the 63% estimated for London in 2010 (Smith, 2010). However, decisions can also be positive, in that landscaping choices can enhance ecosystem services and provide habitat for wildlife.

Many studies have identified the influence of landscaping and planting choices on birds and invertebrates. In the Chicago area, Belaire et al. (2014) conducted bird surveys across residential neighbourhoods and found native bird species richness was positively related to the proportion of yards with plants bearing fruits and berries and those with more trees, in particular a balanced mix of evergreen and deciduous trees, but was negatively related to the number of outdoor cats. Backyards were particularly significant as an underestimated biodiversity resource. Because they are not visible from the street, their management is not subject to the same constraints imposed by social norms, and consequently, they support more vegetation types and wildlife resources, such as plants with fruits or berries, and vegetation that will attract birds (Belaire et al., 2016). In Hobart, Tasmania, garden characteristics had a large effect on garden bird assemblages, indicating that garden design and management can favour particular species (Daniels and Kirkpatrick, 2006). Gardens with greater proportions of native plants and with multiple layers of vegetation, including tall trees, tend to attract a wider variety and greater abundance of native species, particularly birds (Burghardt et al., 2009; Lerman and Warren, 2011; van Heezik et al., 2008; Daniels and Kirkpatrick, 2006). When aggregated, groups of gardens with wildlife-friendly vegetation could provide large patches of appropriate habitat and improve connectivity between patches (Goddard et al., 2009). Good habitat for wildlife is also good habitat for children. Gardens with diverse habitats, in the form of rocks, walls, boggy patches, ponds, dense vegetation, trees, long grass, messy areas and hidden spots provide a range of play opportunities for children as well as resources for diverse species (Table 4.1) (Figure 4.8).

Gardens also provide many social and cultural benefits for householders, who value their gardens for a wide variety of reasons (Freeman et al., 2012). The diversity of these benefits is reflected in the wide range of motivations and values driving preferences and decisions, which are influenced in turn by a number of factors, including socio-economic status, ethnicity and neighbourhood social norms (Hope et al., 2003; Luck et al., 2009). In New Zealand, a combination of householder knowledge about nature, age, socio-economic status and education level explained some of the patterns in plant diversity and community composition across gardens. The legacy of previous owners' gardening practices and former land use is an important consideration when identifying drivers of garden plant community composition (Hope et al., 2003; van Heezik et al., 2013, 2014). In many studies, socio-economic factors emerge as being the most influential predictors of vegetation cover (Hope et al., 2003; Luck et al., 2009), leading to cause for concern regarding the equity of the distribution of biodiversity, which has implications for human well-being. Well-being is strongly related to the amount of vegetation and other neighbourhood natural features, such as species richness and abundance of birds (Luck et al., 2011). In our study of children across three cities in New Zealand, we found that although all children from families of different socio-economic status had access to similar types and quality of public green spaces, the biodiversity they encountered in their own neighbourhoods differed, because children from more deprived homes had gardens with poor biodiversity (Hand et al., 2016). These children were exposed to less biodiversity in their neighbourhoods because private gardens were less diverse than in wealthier neighbourhoods.

Neighbourhood norms play a large role in influencing landscaping preferences (Nassauer et al., 2009). Peterson et al. (2012) found that the landscape choices of residents in North Carolina were best predicted by what residents perceived their neighbours preferred (although these perceptions were often incorrect), but choices were also influenced by ethnicity, income and home ownership, with African Americans and high socio-economic status positively related to a preference for lawn-dominated landscaping over native plant gardens. Since vegetation diversity drives avian diversity, it is not surprising that factors such as neighbourhood age and socio-economic status, which influence the composition and volume of garden, street and park vegetation, ultimately predict bird diversity and community composition (Loss et al., 2009; Luck et al., 2012; van Heezik et al., 2013).

The significance of gardens as wildlife habitat and their vulnerability to what has been termed 'the tyranny of small decisions' has prompted development of

Table 4.1 A list of garden features, their associated wildlife and their values for children's play and learning about nature

Feature	*Wildlife value*	*Child value*	*Feature*	*Wildlife value*	*Child value*
Rock/stone piles	Reptiles, invertebrates, amphibians, birds, small mammals	Construction, climbing	Large colourful flowers, nectar-producing plants	Invertebrates, pollinators, flies, hummingbirds, honey eaters	Observation, picking, enjoyment
Walls – different heights	Lichens, moss, algae, flowering plants, birds, reptiles, butterflies, moths, invertebrates	Climbing, ball games	Large plants e.g. flax,* (Figure 4.8)	Native birds	Creativity, manipulation, construction
Hedges	Birds, mammals, reptiles, amphibians, invertebrates	Climbing, exploring, hiding	Fruiting plants	Birds, mammals insects	Observation, eating
Dense vegetation	Invertebrates, birds, mammals, amphibians	Play, hiding, dens	Seeded plants, e.g., thistles, teasels	Birds, invertebrates	Play, enjoyment
Boggy wet places	Swamp plants, frogs, water dwellers, insects, spiders, slugs	Play, observation, education – life cycle	Wood piles, decaying logs	Invertebrates, amphibians, hedgehogs, mice, fungi	Observation, education, fun, education – life cycle
Ponds and water features	Amphibians, invertebrates, birds, mammals	Play, observation, education – life cycle, food webs	Structures; bird feeders, bird bath, nesting boxes, lizard covers, bug hotel, bat box	Birds, reptiles, mammals, invertebrates, amphibians	Observation, education, fun
Trees	Birds, invertebrates, some mammals	Climbing, observation	Compost heap	Invertebrates, mammals, e.g., hedgehogs, rats, birds, worms	Observation, exploration
Long grass	Invertebrates, birds, mammals, reptiles	Climbing, exploring, hiding	Muddy patches	Birds, invertebrates, amphibians	Digging, construction

* Flax is import also for Māori, who use its leaves for weaving.

Figure 4.8 Some plants, such as flax, are important for native species but also provide material for play, as here the boys are fighting with flax stalks.

the concept of a toolbox for garden governance (Dewaelheyns et al., 2016). In this study, a number of barriers were identified reflecting opinions of residents. They included social pressure to have a neat and tidy garden, the desire to express individualism through gardening, a lack of policy, under-valuation of green spaces by developers, a hesitation to intervene in the management of what is private property and the feeling that there was insufficient knowledge and experience to warrant the development of policy and initiatives. However, barriers could be addressed through the provision of information; engaging the public; setting examples, such as pilot projects; and encouraging and exploring potential ideas.

Value of trees

Urban trees, also called urban forests, provide many important social, economic and environmental services (reviewed in Millward and Sabir, 2011). By shading and through transpiration and evaporative cooling, they prevent warming by intercepting and reflecting sunlight, and they cool buildings and paved surfaces. Vegetation close to buildings can lower indoor air temperatures by as much as 3 degrees C compared to the unshaded areas of the same building. In winter, trees reduce heating costs and act as windbreaks. Trees also reduce air pollution by absorbing pollutants, intercepting particulate matter and releasing oxygen.

Urban forests also act as sinks for carbon dioxide. In northern New England, Zheng et al. (2013) calculated that urban forests sequestered 603,200 tonnes carbon/year ($US38.7 million/yr value), contributing 8.2% to the region's net forest ecosystem carbon sequestration, and Nowak and Crane (2002) estimated that urban trees in the US store 700 million tonnes of carbon. Trees also reduce the amount of storm-water runoff by intercepting rain as it falls, reducing the volume of runoff and the total flow (Millward and Sabir, 2011). For example, in Santa Monica, Xiao and McPherson (2002) showed that the city's 129,000 street and park trees intercepted 1.6% of the total rainfall, resulting in 193,168 m^3 of mitigated storm-water runoff.

Trees can be found in public spaces, such as parks, nature reserves and housing estates and along streets, but a large proportion of the urban forest is made up of trees in private yards. Even a highly developed and compact city like Hong Kong is able to accommodate a relatively rich flora of about 390 tree species, with more than half of these on public housing estates (Zhang and Jim, 2014a). Typically, trees in public spaces and private gardens are dominated by a small number of exotic species, although there is scope for native species with appropriate form and attractive flowers to be given preference in planting programmes (Zhang and Jim, 2014a, 2014b). Despite their non-native status, exotic trees do provide resources for some native species (Gray and van Heezik, 2015); however, native trees can provide a greater sense of identity or place.

Some cities have adopted strategies to actively enhance their urban forest. In Melbourne, Australia, an Urban Forest Strategy was implemented to combat a steady decline in trees following a 13-year drought and aims at 3,000 new trees each year. The city assigned all of Melbourne's 77,000 trees individual emails so that citizens could report on vandalism or declines in tree health; however, in an unexpected turn of events, many people used these addresses to write love letters to trees. For example,

> **Golden Elm, Tree ID 1028612**
>
> I used to think you were the Magic Faraway tree when I was a child.
> Now that I'm an adult, I still look forward to seeing you as I come around the bend after a tedious crawl down Hoddle Street.
> A loyal friend always there waiting to say hello
>
> (News.com.au).

Trees are usually valued by people, and their presence improves the aesthetics of neighbourhoods (Figure 4.9). While trees may sometimes be imbued with great cultural or historical significance and treed spaces provide recreational opportunities, they also provide important habitat for wildlife, particularly when they are large (Chen and Jim, 2008; McPherson and Simpson, 2002; Nowak and Dwyer, 2000; Stagoll et al., 2012). Parks containing greater tree cover or remnant vegetation attract people who appreciate nature (Shanahan et al., 2015), and house prices tend to be higher in neighbourhoods close to well-treed areas, such as parks. In Australia, Kirkpatrick et al. (2012) classified residents according to their motivations for tree planting and tree removal into aesthetes, spiritual tree lovers, practical tree lovers, arboriphobes, native wildlife lovers, tree hazard minimisers and

Figure 4.9 In addition to ecological values, trees also have considerable play value for children. Photo by Ursula Ellenberg

indifferents. They found membership in these categories was predicted by income, tertiary education and gender. Trees can elicit very strong emotions, with a proportion of the public perceived as tree-haters. Conflict over urban trees is not uncommon. The removal of particular trees for developments can be met with passionate resistance by locals, who may try to camp in the trees to prevent their destruction. In Australia, Kirkpatrick et al. (2013) determined the main motivation for the removal of healthy trees is the desire to improve aesthetics and functionality, but in a human population that has widely varying tastes and moral perceptions, in combination with high property turnover rates, old urban trees are highly likely to be removed. In the US, homeowners valued trees on their properties only if the treed areas appeared sufficiently tidy and appropriate for the appearance of the neighbourhood. If more trees were incorporated into the design of sub-divisions at an early stage, acceptance of gardens with trees might be greater, enhancing carbon storage and ecological benefits (Visscher et al., 2016).

Nonetheless, trees are often the first to go when urban developments proceed, as their value is not always well understood or appreciated by planners and developers. Recent studies demonstrate methodologies by which the functional values of trees can be calculated. For example, in the City of Toronto, the 309 trees of 43 species in Allen Gardens, a forested public park in downtown Toronto, provided

USD$26,326 in annual benefits ($16,665 environmental and $9,661 aesthetic), delivering a benefit-to-cost ratio of 3.4:1 (Millward and Sabir, 2011).

Value of birds

How much do people value birds? This is a difficult thing to measure, but because birds are usually the most visible animals in urban areas, they present the best opportunities for urban residents to interact with wildlife. The very high incidence of bird feeding, already discussed, indicates that birds are valued highly (Figure 4.10). However, the high rates of cat ownership in many countries with consequent impacts on wildlife, including birds (van Heezik et al., 2010), suggest people value their cats more than they value being able to see a wide variety and high abundance of birds. In fact, these values are interconnected: a Chicago study found that residents with cats outdoors were also more likely to have a bird feeder, birdhouse or plants intended to attract birds. Outdoor cat owners report a greater fondness for animals than do people who do not have pets, which may explain why they want to attract birds, and they probably believe that their cats do not harm wildlife or are just indulging in natural behaviour (Belaire et al., 2016).

Figure 4.10 Children tend not to differentiate between bird species in the same way that adults do in ascribing values to rare or common birds. Children place a high value on birds that are fun and interactive as with these common pigeons.

Provision of nest boxes and water sources are other means by which people can attract birds. Clucas et al. (2015) attempted to calculate a monetary value for birds by evaluating bird food expenditure, the activities engaged in to attract birds and willingness-to-pay for publicly financed action to increase bird abundance in both Seattle and Berlin. They found that the lower bounds for total economic value placed on enjoyment of songbirds (120 million USD/year and 70 million USD/year for Seattle and Berlin respectively) indicated residents valued birds highly, although in Seattle people preferred to spend their money on attracting birds to their homes rather than on public schemes to improve bird abundance. Willingness-to-pay to attract birds also varied between bird species, with finches being more popular than crows. The observation that older adults spent more on bird food and engaged in more bird activities suggests urban birds provide an important opportunity for older people to interact with wildlife. In general, considerable value was generated by common backyard bird species, both economic and ecological (e.g., seed dispersal, reduction of insect pests).

Optimal landscape design paradigms for ecosystems?

How should we best design our growing cities to preserve the ecosystem goods and services we depend upon? The ongoing movement of people into urban areas and the huge predicted increase in human population growth presage an enormous expansion in the built environment. Net human population growth is taking place mostly within urban areas, which will have to accommodate an additional 2.6 billion people by the year 2050 (United Nations Population Division, 2011). Within cities, the density of residential development can vary hugely, from low-density residential plots surrounded by vegetated yards to high-density housing estates with virtually no remaining vegetation (Fuller and Gaston, 2009; Jenks and Jones, 2010). Urban sprawl, a consequence of low-density development, creates environments with moderately degraded biodiversity over large areas. It results in a loss of peri-urban green space (termed 'land sharing'; Lin and Fuller, 2013), whereas high-density developments have a severe impact on local biodiversity but occupy a smaller area and free up land for large natural reserves (termed 'land sparing'; Lin and Fuller, 2013). However, higher-density development means less access to biodiversity for children (see Chapter 10). Given the reluctance of parents to allow their children to roam independently and widely (we measured a mean maximum distance moved from home of about 500 m; Hand et al., 2016), it is unlikely that children living in high-density developments would regularly visit the large nature reserves enabled by this kind of landscape design (Figure 4.11a and 4.11b).

Sushinsky et al. (2013) modelled the consequences of compact and sprawling urban growth patterns for bird populations in Brisbane, Australia, a city experiencing rapid population growth but with an historical pattern of low-density, sprawling development. They found that compact development involving high-quality large green spaces was better at preserving bird communities and reducing local extinctions, particularly for those species that are already restricted to large green spaces or patches of remnant habitat. However, responses differed between

Figure 4.11a This playground was identified as the most used outdoor area by children attending one school in our *Natural Neighbourhoods for City Children* study. Its popularity was due to its close proximity to the children's homes, but from an ecological viewpoint, it has poor biodiversity value.

Figure 4.11b In contrast, this site with greater ecological value was little used by children outside of organised school visits.

different groups of birds and between species; non-native species were predicted to increase their distributions under sprawling development and were mostly unaffected by compact growth, whereas urban-adapted native species were relatively unaffected by either form of development. The authors suggest that despite the role private gardens play in supporting biodiversity, reducing their size might be a worthwhile strategy for bird conservation if higher-density development with larger greenspaces was achieved. The same question has also been applied to bats, which are negatively affected by urbanisation through loss of roosting habitat (large hollow trees), altered noise and light regimes, changes in prey availability and alterations in interactions, such as competition and predation (reviewed in Russo and Ancillotto, 2014). Housing density in Melbourne, Australia, was the factor that most strongly predicted bat distributions and bat activity. The activity of urban-adapted bats was reduced by more than half once housing density reached only three dwellings per hectare and to almost zero at densities of eight to 10 dwellings per hectare, with only two species able to exploit the urban matrix, apparently unaffected by increasing urbanisation (Caryl et al., 2016). It is therefore unlikely that any land-sharing development would provide suitable habitat requirements for bats, as these dwelling densities are very low, and a land-sparing strategy providing forested habitat remaining in at least 20% of the landscape may be a better option for bats.

Land sparing as a preferred development strategy requires smaller or non-existent gardens. In fact, gardens are already reducing in size in Australia (Hall, 2008). If the clear social benefits provided by private gardens are taken into account, the land-sparing versus land-sharing debate becomes difficult to resolve. Gardens play an important role in enhancing physical and mental health and well-being, as an expression of ownership and identity, as sites for social relationships and for connecting with nature and as sites of domestic produce production (Freeman et al., 2012). Importantly, they are also where children are most likely to encounter and interact with nature (Hand et al., 2016). Childhood daily contact with nature has been shown to be important in determining adult emotional connectedness with nature and perception of neighbourhood nature (Soga et al., 2016). Therefore, limiting children's opportunities to connect with nature in their own gardens is likely to result in a decline in public appreciation of the natural world, potentially creating an important new obstacle to addressing global conservation issues.

Integrating the ecological and human dimensions

Urban ecologists see nature in cities through a lens that is very much focused on the requirements of other species, the integrity of habitats and systems and the potential for enhancing and restoring biodiversity to often small spaces in meaningful ways. Urban parks and other recreational green spaces are typically managed for people to enjoy, and this may come at the expense of their role as places where other species can thrive. Urban ecologists, probably more than any other discipline in biology, also recognise the significance and the potential of the human dimension. While green space provision is vital for human health and well-being, it is important from an ecological perspective that the natural spaces created support healthy ecosystems, rather than travesties of natural systems that are biologically depauperate. It is unfortunate and a reflection of an increasingly urbanised lifestyle that many people do not know what natural habitats look like or do not feel comfortable in wild spaces which lack familiarity. The health of cities, in terms of the ecosystem services and the quality of the nature experiences they provide to adults and children alike, depends on how much communities value and know about nature. Values drive community demands for green spaces and the design of these spaces. The quality of the nature spaces we create in our cities will therefore determine what children understand about nature, how comfortable they feel in nature and the quality of the nature they expect as normal in their immediate neighbourhoods.

References

Allinson, D., Irvine, K.N., Edmonson, J.L., Tiwary, A., Hill, G., Morris, J., Bell, M., Davies, Z.G., Firth, S.K., Fisher, J., Gaston, K.J., Leake, J.R., McHugh, N., Namdeo, A., Rylatt, M., and Lomas, K. 2016. Measurement and analysis of household carbon: The case of a UK city. *Applied Energy* 164: 871–881.

Andersson, E., Barthel, S., and Ahrné, K. 2007. Measuring social-ecological dynamics behind the generation of ecosystem services. *Ecological Applications* 17(5): 1267–1278.

Belaire, J.A., Whelan, C.J., and Minor, E.S. 2014. Having our yards and sharing them too: The collective effects of yards on native bird species in an urban landscape. *Ecological Applications* 24(8): 2132–2143.

Belaire, J.A., Westphal, L.M., and Minor, E.S. 2016. Different social drivers, including perceptions of urban wildlife, explain the ecological resources in residential landscapes. *Landscape Ecology* 31: 401–413.

Burghardt, K.T., Tallamy, D.W., and Shriver, W.G. 2009. Impact of native plants on bird and butterfly biodiversity in suburban landscapes. *Conservation Biology* 23(1): 219–224.

Cameron, R.W.F., Blanuša, T., Taylor, J.E., Salisbury, A., Halstead, A.J., Henricot, B., and Thompson, K. 2012. The domestic garden: Its contribution to urban green infrastructure. *Urban Forestry & Urban Greening* 11(2): 129–137. DOI: 10.1016/j.ufug.2012.01.002

Caryl, F.M., Lumsden, L.F., van der Ree, R., and Wintle, B.A. 2016. Functional responses of insectivorous bats to increasing house density support 'land-sparing' rather than 'land-sharing' urban growth strategies. *Journal of Applied Ecology* 53: 191–201.

Chamberlain, D.E., Vickery, J.A., Glue, D.E., Robinson, R.A., Conway, G.J., Woodburn, R.J.W., and Cannon, A.R. 2005. Annual and seasonal trends in the use of garden feeders by birds in winter. *Ibis* 147: 563–575.

Chen, W.Y., and Jim, C.Y. 2008. Assessment and valuation of the ecosystem services provided by urban forests. In M.M. Carreiro, Y.-C. Song, and J. Wu (Eds.), *Ecology, Planning, and Management of Urban Forests: International Perspectives*, Springer, New York, pp. 53–83.

Clucas, B., Rabotyagov, S., and Marzluff, J.M. 2015. How much is that birdie in my backyard? A cross-continental economic valuation of native urban songbirds. *Urban Ecosystems* 18: 251–266.

Colding, J., Lundberg, J., and Folke, C. 2006. Incorporating green area user groups in urban ecosystem management. *Ambio* 35(5): 237–244.

Cox, D.T., and Gaston, K.J. 2016. Urban bird feeding: Connecting people with nature. *PLoS ONE* 11(7): e0158717. DOI: 10.1371/journal.pone.0158717

Daniels, G.D., and Kirkpatrick, J.B. 2006. Does variation in garden characteristics influence the conservation of birds in suburbia? *Biological Conservation* 133: 326–335.

Davies, L., Kwiatkowski, L., Gaston, K.J., Beck, H., Brett, H., Batty, M., Scholes, L., Wade, R., Sheate, W.R., Sadler, J., Perino, G., Andrews, B., Kontoleon, A., Bateman, I., and Harris, J.A. 2011. *Urban: UKNational Ecosystem Assessment: Technical Report*, UNEP-WCMC, Cambridge, pp.361–410.

Davies, Z.G., Fuller, R.A., Dallimer, M., Loram, A., and Gaston, K.J. 2012. Household factors influencing participation in bird feeding activity: A national scale analysis. *PLoS ONE* 7(6): 339692.

Davies, Z.G., Fuller, R.A., Loram, A., Irvine, K.N., Sims, V., and Gaston, K.J. 2009. A national scale inventory of resource provision for biodiversity within domestic gardens. *Biological Conservation* 142: 761–771.

Dewaelheyns, V., Kerselaers, E., and Rogge, E. 2016. A toolbox for garden governance. *Land Use Policy* 51: 191–205.

Elmqvist, T., Setälä, H., Handel, S.N., van der Ploeg, S., Aronson, J., Blignaut, J.N., Nowak, D.N., Kronenberg, J., and de Groot, R. 2015. Benefits of restoring ecosystem services in urban areas. *Current Opinion in Environmental Sustainability* 4: 101–108.

Faeth, S.H., Bang, C., and Saari, S. 2011. Urban biodiversity: Patterns and mechanisms. *Annals of the New York Academy of Sciences* 1223: 69–81. DOI: 10.1111/j.1749-6632.2010.05925.x

Fernández-Jurici, E., and Tellería, J.L. 2000. Effects of human disturbance on spatial and temporal feeding patterns of Blackbird *Turdus merula* in urban parks in Madrid, Spain. *Bird Study* 47(1): 13–21. DOI: 10.1080/00063650009461156

Forrest, A., and St. Clair, C. 2006. Effects of dog leash laws and habitat type on avian and small mammal communities in urban parks. *Urban Ecosystems* 9: 51–66.

Freeman, C., Dickinson, K.J.M., Porter, S., and van Heezik, Y. 2012. 'My garden is an expression of me': Exploring householders' relationships with their gardens. *Journal of Environmental Psychology* 32: 135–143.

Fuller, R.A., and Gaston, K.J. 2009. The scaling of green space coverage in European cities. *Biology Letters* 5: 352–355.

Fuller, R.A., Warren, P.H., Armsworth, P.R., Barbosa, O., and Gaston, K.G. 2008. Garden bird feeding predicts the structure of urban avian assemblages. *Diversity and Distributions* 14: 131–137.

Galbraith, J.A., Beggs, J.R., Jones, D.N., McNaughton, E.J., Krull, C.R., and Stanley, M.C. 2014. Risks and drivers of wild bird feeding in urban areas of New Zealand. *Biological Conservation* 180: 64–74.

Galbraith, J.A., Beggs, J.R., Jones, D.N., and Stanley, M.C. 2015. Supplementary feeding restructures urban bird communities. *Proceedings of the National Academy of Sciences* 112(20): E2648–E2657. DOI: 10.1073/pnas.1501489112

Gaston, K.J., Fuller, R.A., Loram, A., MacDonald, C., Power, S., Dempsey, N. 2007. Urban domestic gardens (XI): Variation in urban wildlife gardening in the United Kingdom. *Biodiversity and Conservation* 16: 3227–3238. DOI 10.1007/s10531-007-9174-6

Gaston, K.J., Ávila-Jiménez, M.L., and Edmonson, J.L. 2013. Managing urban ecosystems for goods and services. *Journal of Applied Ecology* 50(4): 830–840. DOI: 10.1111/1365-2664.12087

Gill, S.E., Handley, J.F., Ennos, A.R., and Pauleit, S. 2007. Adapting cities for climate change: The role of green infrastructure. *Built Environment* 33: 115–133.

Goddard, M.A., Dougill, A.J., and Benton, T.G. 2009. Scaling up from gardens: Biodiversity conservation in urban environments. *Trends in Ecology and Evolution* 25(2): 90–98.

Goddard, M.A., Dougill, A.J., and Benton, T.G. 2013. Why garden for wildlife? Social and ecological drivers, motivations and barriers for biodiversity management in residential landscapes. *Ecological Economics* 86: 258–273.

Gray, E.R., and van Heezik, Y. 2015. Exotic trees can sustain native birds in urban woodland. *Urban Ecosystems* 19(1): 315–329. DOI: 10.1007/s11252-015-0493-1

Hall, T. 2008. Where have all the gardens gone? *Australian Planner* 45(1): 30–37.

Hand, K., Freeman, C., Seddon, P.J., Stein, A., and van Heezik, Y. 2016. A novel method for fine-scale biodiversity assessment and prediction across diverse urban landscapes. *Landscape and Urban Planning* 151: 33–44.

Hope, D., Gries, C., Zhu, W., Fagan, W.F., Redman, C.L., Grimm, N.B., Nelson, A.L., Martin, C., and Kinzig, A. 2003. Socioeconomics drive urban plant diversity. *PNAS* 100(15). DOI: 10.1073/pnas.1537557100

Howard, P., and Jones, D.N. 2004. A qualitative study of wildlife feeding in south-east Queensland. In S.K. Burger and D. Lunney (Eds.), *Urban Wildlife: More Than Meets the Eye*, R. Zool. Soc. NSW, Sydney, pp. 55–62.

Hunter, M.C.R., and Brown, D.G. 2012. Spatial contagion: Gardening along the street in residential neighbourhoods. *Landscape and Urban Planning* 105: 407–416.

Jenks, M., and Jones, C. 2010. *Dimensions of the Sustainable City*. Springer, London.

Jo, H.-K., andMcPherson, E.G.1995. Carbon storage and flux in urban residential greenspace. *Journal of Environmental Management*45: 109–133.

Jones, D.N. 2011. An appetite for connection: Why we need to understand the effect and value of feeding wild birds. *Emu* 111: i–vii.

Kinzig, A.P., Warren, P., Martin, C., Hope, D., and Katti, M. 2005. The effects of human socio-economic status and cultural characteristics on urban patterns of biodiversity. *Ecology and Society* 10(1): 23.

Kirkpatrick, J.B., Davison, A., and Daniels, G.D. 2012. Resident attitudes towards trees influence the planting and removal of different types of trees in eastern Australian cities. *Landscape and Urban Planning* 107: 147–158.

Kirkpatrick, J.B., Davison, A., and Daniels, G.D. 2013. Sinners, scapegoats or fashion victims? Understanding the deaths of trees in the green city. *Geoforum* 48: 165–176.

Kummer, J.A., and Bayne, E.M. 2015. Bird feeders and their effects on bird-window collisions at residential houses. *Avian Conservation & Ecology* 10(2). http://dx.doi.org/10.5751/ACE-00787-100206

Lafferty, K.D. 2001. Birds at a southern California beach: Seasonality, habitat use and disturbance by human activity. *Biodiversity and Conservation* 10: 1949–1962.

Le Louarn, M., Couillens, B., Deschamps-Cottin, M., and Clergeau, P. 2016. Interference competition between an invasive parakeet and native bird species at feeding sites. *Journal of Ethology* 34(3): 291–298. DOI: 10.1007/s10164-016-0474-8

Lepczyk, C., Mertig, A.G., and Liu, J. 2004. Assessing landowner activities related to birds across rural-urban landscapes. *Environmental Management* 33(1): 110–125. DOI: 10.1007/s00267-003-0036-z

Lerman, S.B., and Warren, P.S. 2011. The conservation value of residential yards: Linking birds and people. *Ecological Applications* 21(4): 1327–1339.

Lin, B.B., and Fuller, R.A. 2013. Sharing or sparing? How should we grow the world's cities? *Journal of Applied Ecology* 50: 1161–1168.

Lindemann-Matthies, P., and Marty, T. 2013. Does ecological gardening increase species richness and aesthetic quality of a garden? *Biological Conservation* 159: 37–44.

Loram, A., Tratalos, J., Warren, P.H., and Gaston, K.J. 2007. Urban domestic gardens (X): The extent & structure of the resource in five major cities. *Landscape Ecology* 22: 601–615.

Loss, S.R., Ruiz, M.O., and Brawn, J.D. 2009. Relationships between avian diversity, neighbourhood age, income and environmental characteristics of an urban landscape. *Biological Conservation* 142: 2578–2585.

Luck, G.W., Davidson, P., Boxall, D., and Smallbone, L. 2011. Relations between urban bird and plant communities and human well-being and connection to nature. *Conservation Biology* 25(4): 816–826.

Luck, G.W., Smallbone, L.T., and O'Brien, R. 2009. Socio-economics and vegetation changes in urban ecosystems: Patterns in space and time. *Ecosystems* 12: 604–620.

Luck, G.W., Smallbone, L.T., and Sheffield, K.J. 2012. Environmental and socio-economic factors related to urban bird communities. *Austral Ecology* 38(1): 111–120. DOI: 10.1111/j.1442-9993.2012.02383.x

MacGregor-Fors, I., and Schondube, J.E. 2011. Gray vs. green urbanization: Relative importance of urban features for urban bird communities. *Basic and Applied Ecology* 12(4): 372–381.

Mathieu, R., Freeman, C., and Aryal, J. 2007. Mapping private gardens in urban areas using object-oriented techniques and very high-resolution satellite imagery. *Landscape and Urban Planning* 81(3): 179–192.

Matteson, K.C., and Langellotto, G.A. 2010. Determinates of inner city butterfly and bee species richness. *Urban Ecosystems* 13(3): 333–347.

McPherson, E.G., and Simpson, J.R. 2002. A comparison of municipal forest benefits and costs in Modesto and Santa Monica, California, USA. *Urban Forestry and Urban Greening* 1: 61–74.

Millennium Ecosystem Assessment. 2005. *Ecosystems and Human Wellbeing: Synthesis*, World resources Institute, Washington, DC.

Millward, A.A., and Sabir, S. 2011. Benefits of a forested urban park: What is the value of Allan Gardens to the city of Toronto, Canada? *Landscape and Urban Planning* 100(3): 177–188.

Nassauer, J.I., Wang, Z., and Dayrell, E. 2009. What will the neighbours think? Cultural norms and ecological design. *Landscape and Urban Planning* 92(3–4): 282–292.

News.com.au. *People around the World Have Been Emailing Trees in Melbourne to Confess Their Love*. www.news.com.au/technology/environment/conservation/people-around-the-world-have-been-emailing-trees-in-melbourne-to-confess-their-love/news-story

Nowak, D.J., and Crane, D.E. 2002. Carbon storage and sequestration by urban trees in the USA. *Environment Pollution* 116: 381–389.

Nowak, D.J., and Dwyer, J.F. 2000. Understanding the benefits and costs of urban forest ecosystems. In J.E. Kuser (Ed.), *Handbook of Urban and Community Forestry in the Northeast*, Kluwer Academic, New York, pp. 11–25.

Orros, M.E., and Fellowes, M.D.E. 2015. Widespread supplementary feeding in domestic gardens explains the return of reintroduced Red Kites *Milvus milvus* to an urban area. *Ibis* 157: 230–238.

Orros, M.E., Thomas, R.L., Holloway, G.J., and Fellowes, M.D.E. 2014. Supplementary feeding of wild birds indirectly affects ground beetle populations in suburban gardens. *Urban Ecosystems* 18(2): 465–475. DOI: 10.1007/s11252-014-0404-x

Perry, T., and Nawaz, R. 2008. An investigation into the extent and impacts of hard surfacing of domestic gardens in an area of Leeds, United Kingdom. *Landscape and Urban Planning* 86(1): 1–13.

Peterson, M.N., Thurmond, B., Mchale, M., Rodriguez, S., Bondell, H.D., and Cook, M. 2012. Predicting native plant landscaping preferences in urban areas. *Sustainable Cities and Society* 5: 70–76.

Robb, G.N., McDonald, R.A., Chamberlain, D.E., and Bearhop, S. 2008. Food for thought: Supplementary feeding as a driver of ecological change in avian populations. *Frontiers in Ecology and the Environment* 6(9): 476–484.

Rollinson, D.J., O'Leary, R., and Jones, D.N. 2003. The practice of wildlife feeding in suburban Brisbane. *Corella* 27: 52–59.

Russo, D., and Ancillotto, L. 2014. Sensitivity of bats to urbanization: A review. *Mammalian Biology-Zeitschrift für Säugetierkund* 80: 205–212.

Saggese, K., Korner-Nievergelt, F., Slagsvold, T., and Amrhein, V. 2011. Wild bird feeding delays start of dawn singing in the great tit. *Animal Behaviour* 81: 361–365.

Shanahan, D.F., Lin, B.B., Gaston, K.J., Bush, R., and Fuller, R.A. 2015. What is the role of trees and remnant vegetation in attracting people to urban parks? *Landscape Ecology* 30: 153–165.

Smith, C. 2010. *London: Garden City? London Wildlife Trust, Greenspace Information for Greater London*, Greater London Authority, London, p. 12.

Soga, M., Gaston, K.J., Koyanagi, T.F., Kurisu, K., and Hanaki, K. 2016. Urban residents' perceptions of neighbourhood nature: Does the extinction of experience matter? *Biological Conservation* 203: 143–150.

Stagoll, K., Lindenmayer, D.B., Knight, E., Fischer, J., and Manning, A.D. 2012. Large trees are keystone structures in urban parks. *Conservation Letters* 5: 115–122.

Stanley, M.C., Beggs, J.R., Bassett, I.E., Burns, B.R., Dirks, K.N., Jones, D.N., Linklater, W.L., Macinnes-Ng, C., Simcock, R., Souter-Brown, G., Trowsdale, S.A., and Gaston, K.J. 2015. Emerging threats in urban ecosystems: A horizon scanning exercise. *Frontiers in Ecology and the Environment* 13(10): 553–560.

Sushinsky, J.R., Rhodes, J.R., Possingham, H.P., Gill, T.K., and Fuller, R.A. 2013. How should we grow cities to minimize their biodiversity impacts? *Global Change Biology* 19: 401–410.

Syme, G.J., Shao, Q.X., Po, M., and Campbell, E. 2004. Predicting and understanding home garden water use. *Landscape and Urban Planning* 68: 121–128.

United Nations Population Division. 2011. *World Population Prospects: The 2010 Revision*, United Nations, New York. http://esa.un.org/unpd/wpp/index.htm

United States Fish and Wildlife Service. 2001. *2001 National Survey of Fishing, Hunting, and Wildlife Associated Recreation.* www.fws.gove

Uren, V., Dzidic, P.I., and Bishop, B.J. 2015. Exploring social and cultural norms to promote ecologically sensitive residential garden design. *Landscape and Urban Planning* 137: 76–84.

van Heezik, Y.M., Freeman, C., Porter, S., and Dickinson, K.J.M. 2013. Garden size, householder knowledge and socio-economic status influence plant and bird diversity at the scale of individual gardens. *Ecosystems* 16(8): 1442–1454. DOI: 10.1007/S10021-013-9694-8

van Heezik, Y.M., Freeman, C., Porter, S., and Dickinson, K.J.M. 2014. Native and exotic woody vegetation communities in domestic gardens in relation to social and environmental factors. *Ecology & Society* 19(4): 17. DOI: 10.5751/ES-06978-190417

van Heezik, Y., Smyth, A., Adams, A., and Gordon, J. 2010. Do domestic cats impose an unsustainable harvest on urban bird populations? *Biological Conservation* 143: 121–130.

van Heezik, Y., Smyth, A., and Mathieu, R. 2008. Diversity of native and exotic birds across an urban gradient in a New Zealand city. *Landscape and Urban Planning* 87: 223–232.

Visscher, R.S., Nassauer, J.I., and Marshall, L.L. 2016. Homeowner preferences for wooded front yards and backyards: Implications for carbon storage. *Landscape and Urban Planning* 146: 1–10.

Warhurst, J.R., Parks, K.E., McCulloch, L., and Hudson, M.D. 2014. Front gardens to car parks: Changes in garden permeability and effects on flood regulation. *Science of the Total Environment* 485–486: 329–339.

White, J.G., Antos, M.J., Fitzsimons, J.A., and Palmer, G.C. 2005. Non-uniform bird assemblages in urban environments: The influence of streetscape vegetation. *Landscape and Urban Planning* 71: 123–135.

Wood, B.C., and Pullin, A.S. 2002. Persistence of species in a fragmented urban landscape: The importance of dispersal ability and habitat availability for grassland butterflies. *Biodiversity and Conservation* 11: 1451–1468.

Xiao, Q., and McPherson, E.G. 2002. Rainfall interception by Santa Monica's urban forest. *Urban Ecosystems* 6(4): 291–302.

Zhang, H., and Jim, C.Y. 2014a. Contributions of landscape trees in public housing estates to urban biodiversity in Hong Kong. *Urban Forestry & Urban Greening* 13: 272–284.

Zhang, H., and Jim, C.Y. 2014b. Species diversity and performance assessment of trees in domestic gardens. *Landscape and Urban Planning* 128: 23–34.

Zheng, D., Ducey, M.J., and Heath, L.S. 2013. Assessing net carbon sequestration on urban and community forests of northern New England, USA. *Urban Forestry & Urban Greening* 12: 61–68.

5 The changing city

The changing urban form

In this chapter, the focus is on the urban setting, its character, its form and how it has changed and is continuing to change. We ask what this means for the growing numbers of children whose lives will be lived in these expanding and transforming urban environments. We look at the changes that are taking place in the built environment and their relationship to green and vegetated spaces. For children, the place that they live is immensely important in supporting or hindering their ability to engage with the natural world. While they may have other experiences where they journey to natural and wilder places in the countryside, it is the environment at the neighbourhood and city level that shapes their most frequent and arguably their most enduring nature relationships. As Wells and Lekies (2006) found, childhood interactions and experiences with the natural world are important determinants of how children will relate to the natural world as adults. We explore the form of the urban environment that children live in. This form matters as more children are being born into and live their lives in urban areas, which 'account for around 60 per cent of urban growth' (United Nations Children's Fund (UNICEF), 2012). The percentage of the population that is urban for a range of countries illustrates this development: Fiji 53%, NZ 86%, Australia 89%, Germany 75%, Japan 93%, Philippines 43%, South Africa 64%, UK 82%, China 54%, Brazil 85% and Iran 73%, with some countries, such as Singapore, being completely urbanised (World Bank, 2016). The proportion of the population living in urban areas is increasing globally, and most children's lives will be urban ones. It is essential that we understand the spatial context that is emerging to understand how nature both fits into and is excluded from children's living environments.

Industrialising and greying cityscapes

Cities have traditionally not been green. Historically, public spaces comprised plazas, city squares, marketplaces and streets. Landscaping was not part of the provisional equation. Green spaces were largely private gardens with access limited to wealthy owners (Figure 5.1), or they were relict habitats, such as

Figure 5.1 Green squares, such as this one in London, whose use was once restricted to surrounding building owners, have increasingly become publicly accessible parks. On a sunny summer Friday evening, this one was being well used by local families.

woodland, encapsulated by the growing city. The generally smaller size of the early cities did mean that the countryside was closer and thus more accessible. In the earlier parts of the twentieth century, country walks were a typical family event, especially on Sundays, and were one way that city children could experience nature. In 'Callum Brown's chapter on the 'Everyday Scottish Sunday', he describes recollections from older people where walks were one of the few Sunday activities (aside from church activities) permitted to children growing up in Presbyterian Scotland (Brown, 2010). In later years though, when rapid growth and urban sprawl became dominant, countryside access was further away so walks became more difficult and in the later 20th century were often replaced by the Sunday drive to the country.

Centuries of relatively slow growth and change ended with the Industrial Revolution as new cities were created and existing cities grew apace. The Industrial Revolution may have helped make Britain 'great' economically, but it certainly did not make it great environmentally or provide good natural and healthy environments for children (Figure 5.2). Neither was Britain alone in the greying of its living environment, as industrialisation spread its tentacles to cities across the globe, exporting the European industrial urban form. The 1919 Town Planning

Figure 5.2 Housing built during the late industrialising phase of city growth continues to be a nature-deprived living environment.

Conference and Exhibition in New Zealand echoed the fears and hopes of many seeing this spread:

> This is a young country, and it is our duty thus early in its history to endeavour as far as possible to avoid the mistakes which have been made by countries in the old lands of the world, especially in the mother country. There the growth of industries during the last hundred years has tended to the aggregation of population in the cities, in the depletion of the rural districts, the creation of slums . . . and to the existence of vast masses of people to whom the amenities and joys of open-air life are practically unknown. What is more important, amid such an environment a healthy race cannot be reared.
>
> (Hon G. W. Russel in NZ Govt., 1919)

Back in Britain, John Ruskin (1819–1900), the Victorian social commentator, said, 'The measure of any great civilisation is in cities and a measure of a city's greatness is to be found in the quality of its public places, its parks and squares' (UK Parliament, 1999). For many, the reality of life was far removed from the imposing and pleasant environment of parks, squares and public places. The legacy of the 19th-century industrial city presented particular challenges as the following description by Stephen Owen of his boyhood journey into Leeds (Yorkshire, UK) shows:

> As a small boy in the 1950s I went to Leeds almost every week. It was a grim experience. Coming over the edge of the Aire Valley in the bus and looking down into Leeds revealed only a heavy, grayish-brown pall through which few buildings were visible and through which little sunlight ever penetrated except during Bank Holidays when the factories were idle. As the bus came alongside them, most of the buildings were black and streaked with iridescent oil. The air was gritty; there were no trees. Back to back houses were overshadowed with engineering works and cheek by jowl with scrap yards piled with rusting metal. The walk from the bus station to the main shopping streets passed by a slaughterhouse with its smells and sounds of death and dribbles of blood on the pavement. On the way home the bus crawled through fog and the bus seemed always to be full of coughing people. It was inconceivable that towns might be enjoyable places.
>
> (Owen, 1991, p. ix, Figure 5.3)

For the majority of inhabitants in the city, life became increasingly divorced from opportunities for positive contact with the natural world. Rivers became transportation

Figure 5.3 A rehabilitated part of the River Aire, rich in vegetation and very close to the bus station where Stephen Owens ended his childhood journey into Leeds.

highways and repositories for pollution, the compact nature of the housing left little room for gardens, and for most urban dwellers, parks and the countryside were inaccessible with regard to both time and transport. Rickets, a disease associated with slums and lack of sunlight and vitamin D, was rife amongst children. The air also was grim because of the prevalence of smog. The dirtiness of the air was reflected in street and area names such as the 'Black' Country (West Midlands) and even in the song 'Dirty Old Town' (by the Pogues, which was referring to Salford, Manchester).

Planning for green space

Together with feelings of despair in the urbanising cities were feelings of hope expressed in a range of ways. Parallel to industrialisation were the development of early formally designed green spaces and parks. Some were impressive in size and construction, such as the Englischer Garten, Munich, begun in 1790; the Botanic Gardens in Sydney (1816); Tivoli Gardens in Copenhagen (1843); Central Park in New York (1857); Vondelpark in Amsterdam (1865); and Golden Gate Park in San Francisco (1871). In Asia, parks were often associated with temples and palaces and were only later opened for public use or were created by colonial regimes (Figures 5.4 and 5.5). The French constructed many of Ho Chi Min City's parks, and in Seoul, parks associated with the royal palaces or hunting grounds (Seoul Forest) provided important green spaces in the rapidly urbanising city. These parks provided and continue to provide respite from the city. They include lakes, such as the Serpentine (Hyde Park); grand avenues and walkways;

Figure 5.4 Temple grounds in Seoul now open to the public are a drawcard for Seoul school children.

Figure 5.5 A statue in the Singapore Botanic Gardens dedicated to children's play; today far fewer children are seen enjoying such free play.

and even racecourses, as with Longchamps (Bois de Boulogne). They were also characterised by less formal areas associated with more naturalistic planting and farm environments, such as in the Amsterdamse Bos. This 100 ha parkland was created in 1937 on unused turf lands and wetlands (Figure 5.6). In his book *Public Parks and the Enlargement of Towns* (1870), Frederick Law Olmsted (1822–1903) laid out the political and philosophical case for public parks as follows:

> A park fairly well managed near a large town, will surely become the centre of that town. With the determination of location, size, and boundaries should therefore be associated the duty of arranging new trunk routes of communication between it and the distant parts of the town existing and forecast . . . no part of the town should finally be many minutes' walk from some of them and they should be made interesting by a process of planting and decoration.
>
> (in Le Gates and Stout, 1997, p. 343)

Figure 5.6 An urban farm which is part of Amsterdamse Bos and a great attraction for families.

While the early intention may have been primarily to provide parks for the middle and upper classes and to indicate the wealth of the nation, they were still public places. So, while the public, and certainly children, may not have been the intended or preferred users of these city parks, their presence was, nonetheless, tolerated. These parks, which are still prized by urban dwellers and used by children today, owe their origins to the far-sightedness and view of public service expressed by municipal beneficence. In structure, they remain fundamentally unchanged from the days of their inception, comprising a mixture of duck ponds, fountains, trees, shady nooks, flowerbeds, lawns and sports grounds. Some new users have emerged, for example, skateboarders, soccer players and joggers, but the fundamental purpose of providing a place of beauty and relaxation away from the stress of city life remains essentially undiminished. The legacy of these parks is of huge importance for children today, as in many cities, especially in the inner city, they remain one of the few open spaces available to them. They represent rare opportunities for providing access to nature and that all-important nature ritual of 'feeding the ducks'.

The unplanned green space

Nature has never entirely been banished from the city. While parks represent formal urban greening, another equally important process has been the encapsulation of natural spaces. Commons were primarily associated with villages and the countryside, but as towns and cities expanded, some commons, such as Clapham Common, Hampstead Heath and Wimbledon Common in London, came to be encompassed within the urban fabric. Sometimes woodlands, rivers and marshes also resisted the spread of development, becoming encapsulated. In London, Epping Forest, Greenwich Park, Highgate Wood and Richmond Park contain relics of ancient woodlands; downland and heathland relics are retained at Hampstead Heath and wetlands at Rainham Marsh. Seoul Forest in South Korea and historic cemeteries, such as Highgate in London (burial place of Karl Marx) are important wildlife refuges and provide woodland sites. Natural vegetation can be found in neglected spaces or spaces difficult to develop, such as along riverbanks or along steep slopes. Over time, most are lost to or eroded by development, but some remain – either by accident, saved by their unsuitability for development (e.g., marshes and wetlands) or through popular support. Most cities include elements of 'leftover spaces' with often good habitats; particularly common are estuaries, rivers and hillslopes, derelict land, quarries, spoil heaps, docks, sewage works, railway yards and transport corridors, such as rivers/canals, motorway verges, roadsides and railway reserves. As factories close or move to the outskirts, slums are cleared and railway yards and canals become less used for industrial transport and more for leisure, and as polluting industries cease, become cleaner or move to the urban periphery, nature has been able to reclaim these areas as its own. These left-over landscapes, which included post-war bombsites, became playgrounds for generations of children in post–World War II (WWII) Europe:

> As a child, playing outside was an everyday occurrence. There were few cars on the roads, and plenty of amazing pieces of derelict land which grownups referred to as bomb sites. These were our playgrounds. Children were free to wander . . . Clusters of bushes and tangled mounds of rubble and dandelions became our 'homes'.
>
> (Andrews, 2012 p. 97)

Unmanaged, often untouched and unused, these leftover spaces are colonised by a range of both native and exotic plants. These are the landscapes Hough (1989) refers to as the 'non-pedigree landscapes'. Frequently undervalued and treated with disregard, subject to erasure at the whim of developers, they are, nonetheless, often the richest natural areas in the urban fabric. Only with the development of the green city, sustainable development and urban nature movement of the late 1980s has the value of these 'non-pedigree' landscapes gained any positive recognition. The post-war period then saw two parallel developments – the displacement of nature by and to the suburbs as cities intensified their built core, and a much quieter and less planned recolonisation by nature of space surplus to the urban planners' requirements. Nature has colonised and still colonises unused land in the city. Formal green space planning is only part of the process of creating natural spaces in the city. The responses of urban professionals to these naturally colonised spaces are important in determining whether they continue to naturalise or become lost to development. Historically, these natural, usually overlooked spaces have been important playgrounds for children. Some of the key phases in this city greening process are now explored.

Greening the living environment

Garden city to suburb: In the 1900s, the search for utopia continued with a blending of built and urban forms. Ebenezer Howard's Garden City, first described in his 1898 book *To-morrow: A Peaceful Path to Real Reform* (Howard, 1898), is perhaps the best known of the urban visions. Its appeal can be seen in the lasting real estate desirability of homes located in garden suburbs and cities a century later. Like Olmsted, Howard believed passionately in the need to find a better way of urban living than that provided in the overcrowded industrial cities of his time. The way to achieve this, he believed, was to unite town and country: "Human society and the beauty of nature are meant to be enjoyed together". Thus, the notion of the family home with a garden for parental enjoyment and children's play was born, ideas not that dissimilar from those promoted in the sustainable city ideas of the 1990s. The principles these garden cities embodied could be seen in the rise of suburbia, described somewhat poetically by Ward as follows:

> Regarded as a childhood idyll, the suburb is at its best in mid-afternoon. It is quiet and still and the only sounds are the sounds of childhood . . . It is the child rearing sector of the city: its nursery 'for the sake of the children'. The first decision of any family with the luxury of choice has been to opt for suburbia.
>
> (1977, pp. 72 and 73)

Suburbs have played a contradictory role in relation to nature. First, their expansion has swallowed up countryside, creating the sprawling cities so common today, especially in Australia and the US. Secondly, the gardens of these same suburbs, especially the older, usually inner suburbs have in many cases become refuges for many plant and animal species displaced by the intensifying rural agro-economy. Many suburban gardens mimic the woodland edge, which is one of the richest of biotas. The presence of food supplied directly by people has contributed to healthy populations of some species. Suburban gardens provide rich potential for children's direct engagement with nature, whether it be through feeding birds (half the adults in the UK feed birds (Royal Society for the Protection of Birds (RSPB), 2009), climbing trees, growing vegetables or simply outdoor play (Figure 5.7).

High-rise peril: Whilst Howard's vision focused on the small scale, the uniting of the urban and the natural, the vision of Le Corbusier (1887–1965), key architect of the Modernist movement reflected quite different ideas. In his book *Urbanisme* (1925), he described how there were to be no more congested streets and pavements, no more busy public squares and no more untidy neighbourhoods. Instead, people would live in hygienic, regimented high-rise blocks interspersed with park-like landscapes – a rational city that would be separated into discrete working, leisure and residential zones. This vision, at least in Europe, failed through being inhospitable, disorienting, socially destructive and dispiriting. The proposed

Figure 5.7 Suburban gardens provide a range of opportunities to engage with nature; here, boys eat strawberries grown in the garden.

conditions can lead to disaffected residents who have, as in the *quartiers difficiles* or *banlieue* (French suburban high-rise neighbourhoods) near Paris, physically revolted.[1] For children, the legacy of le Corbusier has been disastrous. High-rise estates, whether in France, UK or Eastern Europe (Figure 5.8), have been one of the greatest town-planning failures ever and have been anathema for those who believe that contact with nature is essential to children's well-being. The potential for contact with nature in these high-rise estates is as limited as it was for children condemned to live in the slums of Charles Dickens's 19th-century industrial 'coketown'. The 1990s saw the gradual dismantling of the high-rise in many parts of Europe, but in other parts of the world, it very much remains the dominant residential urban form. For children growing up in places like Hong Kong or Singapore, there is rarely an alternative to high-rises. Yet, Singapore, through its city greening strategy, manages to mimic the garden city ideal through surrounding these high-rise apartments with green spaces that include streams, trees, bushes and plantings of various types and many nearby parks and greenways.

Repopulating the inner city: Until the 1990s, city development was fairly stable – high-density low-rise housing constructed during the late 1800s and 1900s, suburbs gaining precedent through the 1900s and high-density high-rise housing constructed in the 1960s and 1970s followed by its partial dismantling in the 1990s. Developments that are more recent include the repopulation of the inner city,

Figure 5.8 The legacy of le Corbusier style high rise, Plymouth, UK; built in 1964, these apartments remain an alienating environment for children to grow up in.

especially by families, as described in Whitzman and Mizrachi's *Vertical Living Kids* (2009). In their study, they identify a reverse trend of families moving back into the inner city. Inner-city living, when it is in apartment blocks, can be alienating if there is no private garden space in and around the housing block. For these children, it is the public spaces that will be their places of natural encounter, but this can only occur if children are themselves experiencing these green spaces regularly and at length. As cities become greener and parks and other green spaces become increasingly characterized by more natural planting, there is no reason why inner-city living need be the nature-deficit experience characteristic of the inner-city living of last century. As Figures 5.9 and 5.10 show, there is considerable potential for integrating nature even into high-density living environments. The designs were part of an exhibition of urban designs for greener city living created by Singapore high school students and displayed at the City Gallery in 2015. For children living in cities, the range of potential places to encounter nature can be substantial. The key changes in housing styles, the different types of green spaces available and opportunities for both the exclusion and inclusion of nature are highlighted in Table 5.1.

Figure 5.9 Designs for greener living by Singapore school students, ‘Vision integrating nature: let love breathe green’.

Figure 5.10 Designs for greener living by Singapore school students, ‘Green capillary: Forest for rest, living in zest, connecting the mass’.

Table 5.1 Changes in city living environments for children post industrialisation

Key developments in urban form	*Nature values for children*	*Green spaces that can be used by children including child-specific green spaces*
Industrialised city: high-density housing, commonly terrace or row housing, usually small, closely packed houses with little public or private green space	Generally poor – low quality, dominated by hard landscaping	Humanised green space • Private gardens • Courtyards • Allotments, community gardens • Landscaping and flowerbeds • Lawns • Public spaces – cemeteries, hospital grounds, sports fields.
Garden city: low-rise housing with gardens	Rated highly for family living; older suburbs tend to be well vegetated, but newer suburbs, while well used for families, can be somewhat sterile and uninteresting with the trend towards larger homes with smaller gardens	Planned green space • Public parks • Botanic gardens • Green/town belts
Suburbanisation: low-rise housing with gardens, often part of very large cities typical of Australia, North America and parts of Europe	As for the garden city	Natural green space • Woodland • Meadows • Lakes, streams
High-rise: often in a landscaped setting and increasingly the dominant housing in rapidly urbanising Asian nations	Low natural value and alienating from nature with rare exceptions such as in Singapore	Industrial • Derelict land – informal open space • Reclaimed land, pits, industrial land, quarries
De-industrialisation: the repopulation and enhancement of inner-city environments as industries close or move to the city outskirts	Varies depending on presence of existing green spaces but does enhance the potential for natural recolonisation	Infrastructure • Roadsides, footpaths • Disused railway lines • Avenues, boulevards, street trees • Canals, rivers
Compact city: move to build at medium and higher densities to reduce urban sprawl, associated with repopulation of the inner city	With good green space design, can be good provision of nature; compactness frees up more open space, especially combined with rehabilitation of canals and other industrial infrastructure	Child spaces • Playgrounds • School grounds • Family gardens • Environmental education – nature parks • Adventure parks • Streets
City greening and eco cities	Projects have immense potential for enhancing children's ability to easily access natural green spaces	

Some newer initiatives: For cities, there are both positive and negative changes in urban development that impact on children's living environments. More children are living and will continue to live in denser city environments. How can these still provide children with the nature contact they need and want? Next, we briefly look at two examples of denser suburban environments and a high-rise environment. In both examples, a conscious effort has been made to naturalise the city but following a more compact urban form.

Box 5.1 Upton, United Kingdom[2]

Upton, Northampton, is described as "one of the most exciting developments planned in England today. It will be a national example of best practice in sustainable urban growth and a model for the design of other urban extensions across the country". The planning brief for the site states, "Parameters establish the principle for a range of open spaces including playing fields, secondary open spaces with children's play, allotments and informal landscaping'. The higher-density and on-street house frontages encourage children's socialising, almost 'forcing' children into the adjacent green spaces.[3] The lack of large private gardens encourages use of the public spaces and the multipurpose infrastructure-landscaped open spaces also encourage outdoor play. Although not just designed as a child-friendly development, with its compact form, intensive use of green space design, street frontages that encourage social interaction, and higher densities, the development inevitably becomes child-friendly. It uses a range of natural design features including:

- Sustainable urban drainage systems: these also act as natural play spaces (see photo)
- Porous paving in courtyards and home zones in residential areas
- Energy conservation through passive solar gain
- Water conservation: high-efficiency fittings
- Rainwater harvesting technologies
- Materials: recycled or local, sustainably sourced materials
- Waste minimisation
- Green roofs and vegetated roof structures

Historically, the development reflects Howard's garden city ideal of bringing together town and country but also meets the sustainability criteria of environmentally appropriate development.

Box 5.2 Singapore

"We want a Singapore that is beautiful – and prosperous – not only for ourselves, but for our children and grandchildren. We have to think long-term," stated the Singapore Green Plan in 2002 (Ministry for the Environment, 2002, p. 52). Singapore is green despite having the world's second highest population density. It has 350 parks, 0.8 ha of parks per 1,000 residents and 220 km of park connectors that are proposed to increase to 350 km by 2020. Green spaces are distributed across housing areas with the aim of ensuring that at least 85% of residents and their families live within 400m of a park by 2030 (Ministry National Development, MND).

The current 1.2 million housing units, of which 0.9 million are Housing and Development Board flats, are estimated to increase to 1.9 million homes. Given this growth, the government has developed a strong greening plan to ensure development is as 'eco' as possible. The direction of building will be upwards, following the high-rise pattern that dominates Singapore's housing. What does this mean for children's connection to nature? Housing has been and will continue to be planned on a precinct basis of 400 to 800 dwelling units set in a green landscaped area with an inner precinct garden with larger parks serving several such units. There is no private green space (Yuen, 1996). The emphasis on communal green spaces can be seen in the My Tree House, the world's first green library for kids, where 30% of books will be on green topics, and the impressive children's garden in the Botanic Gardens. Further parks are increasingly including more natural features and catering for children. One example is the fishing pond in Pasir Ris Park, which, while artificial, nonetheless, provides children with an important way to get close to nature. Given its challenges and the high-rise housing, the provision of green space close to home for children's play is deeply impressive.

What type of city will children inhabit?

This chapter has focused on the living environments of children and identified some of the major building changes that have occurred over the last century or so. City building forms have not only substantially and permanently changed living environments for children in the west but for children across the urbanising world. With these changes come alterations in the ways that children can access and relate to nature. However, as the last two case studies of Upton, UK, and Singapore show, cities need not be a nature-deficit environment. Some would argue that the nature present in cities and towns such as these two is a compromised, artificial nature and 'not the real thing'. While this argument carries weight, what we have done here is to demonstrate that cities in their growth underwent and some unfortunately still are undergoing a process of denaturalisation, making it difficult for children to access nature in their daily lives. However, while recognising this, it is necessary to debunk the myth of an arcadian nature-infused past for children. For most children growing up in cities historically, this was far from their reality. Cities were deprived of sunlight and natural green spaces, and gardens were the preserve of the rich. While children could see animals even in the city; sheep and cattle going to market, horse transport, free-roaming domestic animals and rodents such as rats, there was also 'overt animal cruelty' and mistreatment. This is detailed in Velten's (2013) book *Beastly London*, where dancing bears, organ grinders' monkeys and other animals gave children unhealthy experiences of relationships with animals. In cities today, especially those in which growth is stable or proceeding at a manageable rate, good planning is achievable. Achieving good planning and providing quality green spaces is far more difficult in cities that are experiencing rapid urbanisation and that are resource-poor and struggle to provide basic amenities as in African cities such as Lagos or Indian cities such as Kolkata.

Cities can provide excellent places for children to grow up and to experience nature in ways not always possible for city children in previous generations. However, for this excellence to be achieved requires a whole-city approach and for children to be able to freely interact with city nature. In 1977, Colin Ward produced his seminal text *Child in the City* with these poignant words:

> I don't want a Childhood City. I want a city where children live in the same world as I do. Because some bit of the city is designated as a play space on a plan there is no guarantee that it will be used as such, nor that other areas will not be. If the claim of children to the city is to be admitted, the whole environment has to be designed and shaped with their needs in mind.
>
> (p. 204)

In this book, we too propose a whole-city approach. A careful look at the fabulous black-and-white photographs of city children in Colin Ward's book shows children out and about all over the city, cycling, catching trains, playing football, waiting, climbing scaffolding, working and go-carting. The photographs overall emanate joy and activity, but in urban landscapes largely devoid of green, and where green is present, it is invariably lawn and street trees. In the nearly 40 years since Ward's book was published, cities have changed markedly, and the green is being nurtured, encouraged and prioritised in ways not present then. However, this greening brings its own challenges.

Moving forward with the built environment

The move towards the creation of green cities requires a rethinking of what nature comprises and how children can engage with nature in this now dominant urban living space. Some green space in cities has little obvious natural value. This is most clearly seen in the vast swathes of single-species mown lawn with a few small ornamental trees scattered here and there. It is a ubiquitous urban landscaping type that pervades nearly all urban environments, especially environments seen as suitable for children. It costs huge amounts to maintain in terms of mowing the grass and applying fertiliser and weed killer, all to support what is often an exotic, poorly acclimatised single-species habitat. Lawns do have benefits, especially for active play, but need to be part of a wider green assembly. Other habitats, in contrast, provide ecologically rich and varied examples of nature coexisting and even thriving in the built environment. As Chapter 10 shows, city greening can be highly successful even in initially unpromising environments. A rehabilitated part of the River Aire in the city centre is shown in Figure 5.3. When Owens was a boy, he would likely have passed over the river on his journey into Leeds. At that time, the river would have been smelly, heavily polluted and devoid of any vegetation or wildlife, without the ecologically rich wetland flora found there today. Like the River Aire, cities can recover, and if cities are to be ecologically diverse and good for children, the following set of natural design principles provide a useful guide on how to aid this recovery:

1. Natural systems should be protected and enhanced in urban areas wherever possible, e.g., accessible stream edges should be preserved for children, rather than steep concrete culverts (Figure 5.11 and 5.12).
2. Support ecosystem functioning, e.g., install penetrable rather than impenetrable surfaces to facilitate ground-water penetration; support rather than drain wetlands, which can be used as informal play spaces (see Upton example).
3. Include diverse forms of green space rather than monocultures, e.g., lawn.
4. Natural spaces are more robust and can be cheaper to manage than highly artificial green spaces and tolerate children's 'mess' more than highly formal landscaping.
5. Green spaces that are easily accessible to and available for children also promote use by other people.
6. Use natural rather than hard materials, e.g., woodchips rather than rubber mats in playgrounds.
7. Planting and landscaping should be as natural as possible, providing more interesting environments that also support more species.
8. The natural and built environment should be planned as one, with green spaces as a formative concern rather than an afterthought.
9. In planning for nature, the other values and functions of green space have to be recognised, e.g., the aesthetic benefits of flower beds and the need for effective flood control.
10. Built environments that include green spaces are healthier.

Figure 5.11 An urban stream that retains a more natural form.

Figure 5.12 An urban stream that has a more artificial form that suits its town centre location; both provide excellent child-friendly environments.

Internationally, the form of cities is changing usually, but not always, for the better. Deindustrialisation provides opportunities to green the city. In many places, inner-city environments are becoming greener and pleasanter and can provide good living environments for children. There is no ideal built form. In Singapore, the high-rise option is mixed with intensive provision of green spaces and parklands. Elsewhere, suburbs continue to provide children with low-density homes set in large private gardens. The new and many traditional developments in the Netherlands feature high-density homes in ecologically supportive public environments in which water is central. Developments can support rather than destroy ecosystems. The irony though is that as cities become greener and developers increasingly recognise the benefits of working with natural ecosystems, children are becoming less likely to access that green, given the widespread trend towards decreased independent mobility and outdoor play. For the city's ecological and green space values to be realised, natural spaces have to actually be used by children.

Notes

1 While the 2005 and other uprisings have complex causes relating to poverty, deprivation and others, the inhumane, estranging built landscape contributes markedly to the disaffection present in the marginalised feelings of the populations living there.

2 Information provided by the following sources and an on-site field visit: www.rudi.net/files/Upton.pdf.

Princes Foundation UPTON, NORTHAMPTON Sustainable Urban Extension; information leaflet, nd.

Creating a sustainable urban extension – a case study of Upton, bestpractice@est.org.uk www.est.org.uk/housingbuildings 2006.

3 On my visit, there were plenty of children out and about in Upton after school.

References

Andrews, J. 2012. *A Day in a Life*. Gloucester: Memoirs Publishing.

Brown, C.G. 2010. Spectacle restraint and the sabbath wars: The everyday Scottish Sunday. In L. Abrams and C.G. Brown (Eds.), *A History of Everyday Life in Twentieth Century Scotland*, Edinburgh: Edinburgh University Press, pp. 153–180.

Hough, M. 1989. *City Form and Natural Process*, Routledge, London.

Howard, E., Sir. 1898. *Tomorrow a Peaceful Path to Real Reform*, S. Sonnenschein.

Le Corbusier. 1925. *Urbanisme*, Les éditions G. Crès et Cie, Paris.

Le Gates, R.T., and Stout, F. (eds.). 1997. *The City Reader*, Routledge, London.

Ministry for the Environment. 2002. *The Singapore Green Plan*, Ministry of the Environment, Singapore.

Ministry for National Development Singapore. *A City in a Garden*. www.mnd.gov.sg/landuseplan/city_in_a_garden.htm accessed 2016.

NZ Government. 1919. Official Volume of Proceedings of the First New Zealand Town-Planning Conference and Exhibition. Wellington, 20th to 23rd May, 1919, M.F. Marks, Government Printer, Wellington.

Owen, S. 1991. *Planning Settlements Naturally*, Packard Publishing, Chichester.

RSPB. 2009. *Feeding Birds*. www.rspb.org.uk/makeahomeforwildlife/advice/helpingbirds/feeding/

UK Parliament. 1999. Select Committee on Environment, Transport and Regional Affairs. *Twentieth Report*. https://publications.parliament.uk/pa/cm199899/cmselect/cmenvtra/477/47706.htm

UNICEF. 2012. *The State of the World's Children 2012 Children in an Urban World United Nations Children's Fund New York, NY 10017, USA*. pubdoc@unicef.org. www.unicef.org

Velten, H. 2013. *Beastly London: A History of Animals in the City*, Reaktion Books, London.

Ward, C. 1977. *The Child in the City*, The Architectural Press Ltd., London.

Wells, N.M., and Lekies, K.S. 2006. Nature and the life course: Pathways from childhood nature experiences to adult environmentalism. *Children Youth and Environments* 16(1): 1–24.

Whitzman, C., and Mizrachi, D. 2009. Vertical Living Kids: Creating Supportive High Rise Environments for Children in Melbourne, Australia. Faculty of Architecture, Building and Planning, University of Melbourne.

World Bank. http://data.worldbank.org/indicator/SP.URB.TOTL.IN.ZS accessed 2016.

Yuen, B. 1996. Creating the garden city: The Singapore experience. *Urban Studies* 33(6): 955–970.

Part III

Children and nature interactions

6 Children, families, nature and the home

Formative encounters

For children, home is the place where they first encounter nature and from where they begin their wider exploration of the natural world. By home environment, we mean not just the actual house or apartment but the space immediately around the house, including private and communal gardens and adjoining street space. The home environment is also where, through their families, children develop ecological values and experience their early connections with nature. Keeping pets, playing in gardens, growing vegetables, finding bugs, raising chickens and feeding birds are some of the formative experiences taking place here. Families also introduce children to the wider natural world through family outings, camping, bush-forest walks, picnics and beach trips. In many parts of the world, working on the family vegetable plot, plantation or family farm either in the urban area or during visits out of town is an important part of home life. As children's independent range diminishes, as is the case for many children in western societies, so too does their ability to seek out more distant natural spaces. Thus, the home environment becomes even more important as a source of nature connection. If children live in homes where nature is lacking, the disconnect with nature is heightened. Children whose lives are constrained by finance; family circumstances; political upheaval, including war; ill health; lack of transport; or other factors may be unable to access more distant natural places and rely heavily on what is provided close to and in the home environment.

In this chapter, we look at how families relate to and create nature connections. We then turn to the home itself and different forms of homes, including new home designs and home-based outdoor spaces. The final section looks at creating homes that are more natural. The chapter reveals the necessity and the potential for nature availability in the child's home environment. This includes both planned nature in the form of trees, vegetable gardens and birds using feeders, as well as incidental nature in the form of species, such as insects, spiders, cockroaches, fleas, moths, rats, mice, snakes, lizards and others who make the child's home also their home. Each location and country will have its own assortment of home-based 'companions'. A child living in Brisbane, Australia, would likely have inside his or her home geckos lurking behind paintings, spiders (some huge, such as the huntsman), cockroaches, possums on and in roofs and various birds and reptiles in the garden. In Trivandrum (Thiruvananthapuram) in India, snakes can be present in the home environment but are not always tolerated in the same way as the gecko might be in Brisbane (Figure 6.1a and 6.1b).[1] The values and attitudes given to these incidental creatures will vary by culture and by family, and it is families that are the first focus for this chapter. Particular attention is given to gardens and the biodiversity in the child's home environment, including indoor fauna and flora, those deliberately cultivated like pot plants and those less so, such as weeds and flies.

Figure 6.1a Reptiles, such as this Asian house gecko, are common inhabitants in homes in hot climates; this one is emerging from behind a painting in a house in Brisbane, Australia.

Figure 6.1b A less fortunate reptile, a snake caught in a household drain in Thiruvananthapuram, Kerala, is pounded to death by the extended family as the children watch and learn to fear snakes and see them as a threat.

Families matter

Family is the conduit through which the young child, especially in the early formative years, experiences the world and is introduced to the natural world and its inhabitants, many of which they will interact with or be aware of from their earliest days. Children's earliest memories are often of natural interactions. For one of the authors, a very early memory is watching her dad rescue her bike from the duck pond where some older lads had thrown it, stroking the rag and bone man's horse, a brother coming home with a cracked head after falling out of a tree and walking to Lilford Park with Grandad to see the deer. These occurrences all took place in the home environment, a northern English city, and as part of everyday life. Her family were not 'nature people', but unintentionally, they gave their children a great gift, growing up in a time when indoor play was not encouraged; the norm was for children to be sent outside to play and amuse themselves only coming home for meals. Outside, she roamed far with confidence in the invincible abilities of older siblings to look after her. As families shrink, there are fewer siblings available. The authors' families have between them nine children, not unusual for families in the 1960s, and it meant there were usually siblings available to go out exploring with. Now, as families have fewer children, move around and have busier lives, children's time and freedom to roam widely has contracted, which means that nature interactions inevitably become more closely bound to the home environment. Free-roaming children are becoming less acceptable, as growing adult supervision becomes the norm. Parental attitudes though have always been important in determining children's nature connection opportunities, as these examples from famous conservationists illustrate. The first is from the well-known British primatologist Jane Goodall, who tells the following story of how she worked out the 'origin of the egg':

> When I was about four-and-a-half I was staying in the country, which is very exciting because we lived in London at that time. And I met cows and pigs and horses for the first time, and I had a job of helping to collect the hens' eggs. Well that was fine. The hens were free and they were supposed to lay their eggs in these little wooden hen houses. And I was popping the eggs into my basket, but there's the egg, so where on the hen did the egg come out? I couldn't see a hole like that. And apparently, I was asking everybody and nobody told me, so I decided I must find out for myself. I remember so well. Okay, four-and-a-half, but I remember seeing a hen. She climbed up this little sloping plank into this house where she would spend the night, but they also had the nest boxes. And I followed. A mistake! Squawks of fear, she flew out. So this was not a safe place for a hen. I went to an empty hen house, hid in the straw at the back and waited, and waited, and waited, and the family had no idea where I was. They were all searching. The dusk was falling. My mother sees this excited little girl rushing towards the house all covered in straw. Instead of getting mad at me, you know, 'How dare you go off without telling anyone?' – which would've killed the excitement – she saw my shining eyes and sat down to hear this wonderful story of how a hen lays an egg. So honestly, if you look at that story with hindsight, that has the makings of a little scientist.
>
> (Academy of Science, 2009)

Although this is a countryside story, it is also the story of a mother who, despite her own obvious concerns, recognises the importance of Jane's natural encounter and supports it.

Similarly, the following story from David Attenborough shows his parents' licence – if only by non-interference – that supported David's own natural inclinations. At age seven, David collected newts and sold them to his father's colleagues at the University of Leicester at 3 pennies a newt. When older, he recounts,

> When I was about 13, I cycled from Leicester to the Lake District and back again, collecting fossils and staying in youth hostels. I was away for three weeks, and my mother and father didn't know where I was. I doubt many parents would let children do that now.
>
> (Lusher, 2014)

A third example where interaction was encouraged – though he grew up to be a politician rather than a conservationist – was that of Theodore Roosevelt. Though his parents are not recorded as encouraging this interest, they clearly did not discourage it. Aged eight, he

> collected everything within his reach and range of vision, and begged friends and family to bring him any specimens they found. He even paid other children to collect specimens for him . . . Theodore travelled with his family to Egypt and Syria, where he collected numerous birds. By then a skilled taxidermist, he skinned and mounted the birds himself.
>
> (*American Experience*, accessed 2-10-2016)

For parents today, such 'hands off' parenting with its encouragement of absence or of pursuits such as taxidermy would be difficult. Encouragement for parents today may have to come in a different form, but parents and families still play an important role. In our *Natural Neighbourhoods for City Children* study, children talked frequently and fondly of walks with parents to the park, the bush, and the duck pond and talked about what they saw there. Growing vegetables and other plants and tending family animals is part of family life. In Fiji, tending the family 'garden' remains a vital part of children's lives even in urbanising Suva, the capital city. Boys, in particular, are expected after school and at weekends to tend the family 'garden', where the main crops grown are cassava and dalo/taro. In the maps they drew of their neighbourhood lives, gardens were commonly featured. While these crop gardens are not normally seen as nature, such interactions enable children to relate to plants, growth and seasonal cycles; the soil and its inhabitants; insects that feed on the crops; weather patterns; and the relationship between people and the land. These gardens were far more likely to feature on the maps of indigenous Fijian boys than maps drawn by girls and Indo Fijian children (see the example in Figure 6.2a). The second map (Figure 6.2b) is full of natural elements. This child lives in housing attached to a tourism resort that supports a rich, biodiverse environment, and the influence of this could perhaps be transferring into this child's unusually green map of Suva. The family environment and family

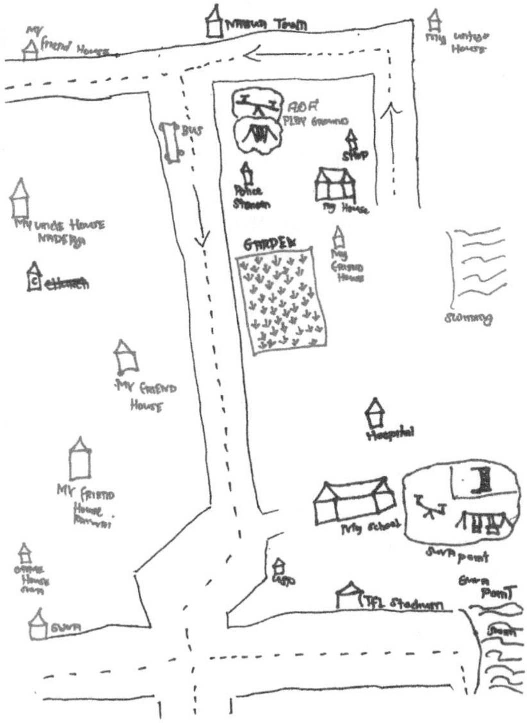

a

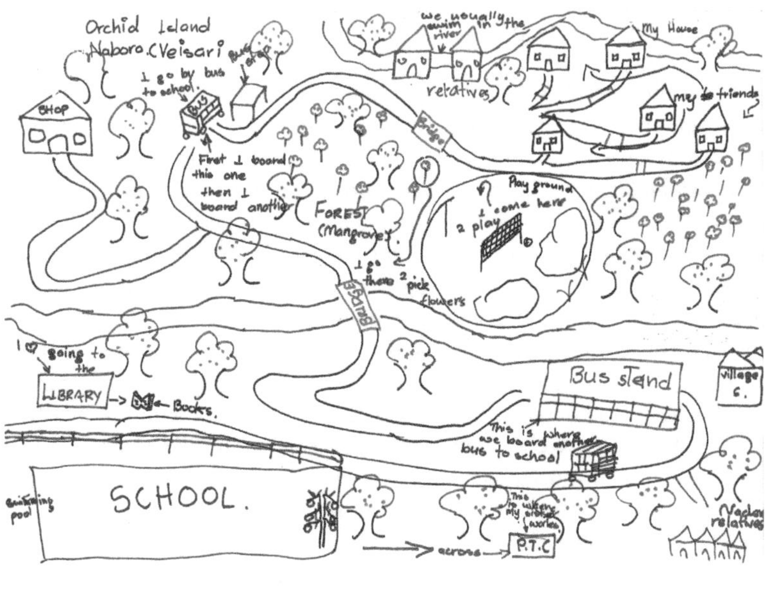

b

Figure 6.2a and *6.2b* Neighbourhood maps drawn by children living in Suva, Fiji's capital city.

expectations, to some extent, regardless of location, impact on how children relate to nature. Important in this respect are family practices, such as holidays and the keeping of animals.

Holidays

The family holiday can be an important way of connecting with nature. The Family Holiday Association's research found that 49% of British people said their happiest memory is being on holiday with their family. Further, when asked their favourite activities, at least two-thirds of the top 15 activities related to the natural environment and associated activities, such as jumping the waves or digging (*Packham Huffington Post*, 2015).

1 Eating ice creams (50%)
2 Collecting shells (42%)
3 Jumping the waves (39%)
4 Building a sandcastle with siblings (38%)
5 Seeing the sea for the first time (36%)
6 Digging a giant hole (33%)
7 Rock pooling (32%)
8 Eating a stick of rock (26%)
9 Being allowed to stay up late (23%)
10 Riding a donkey (21%)
11 Burying another member of my family in the sand (21%)
12 Crabbing[2] (20%)
13 Flying a kite (16%)
14 Fishing (9%)
15 Hunting for treasure (9%)

Holidays provide a basis that helps sustain families' well-being and often prioritise interactions with and spending time in natural environments (Figure 6.3a and 6.3b). A study of family campgrounds in New Zealand estimated that around 80% of New Zealanders have been camping, mostly during childhood (Department of Conservation, 2006). Another New Zealand study found parents adopted more hands-off parenting styles while camping and actively sought out less commercial campsites that offered natural play opportunities, such as swimming, water play, fishing, tree climbing, cycling and general free-play (Freeman and Kearns, 2015). At the campsites, it would be unusual not to let the children run free, and like the British study, families valued the memories the camping provided. In conducting the study, the researchers became aware of an odd contradiction in that freedom and independence were encouraged in a relatively unfamiliar place populated by relatively unknown individuals, whereas at home, most of these same parents placed significant constraints on their children's freedom and independence and limited access to natural play spaces (Freeman and Kearns, 2015). The parents interviewed valued strongly their children's natural experiences on holiday and

a

b

Figure 6.3 Beach activities: (a) Enjoying the beach in Devon, UK, and (b) A Friday night in Thiruvananthapuram, Kerala, when the beaches are packed with families watching the sun go down and children play at the sea's edge.

Figure 6.4 For most children in this photo, going to the beach (East London, South Africa) as part of a preschool trip, this was their first ever beach experience.

simultaneously regretted their loss in the home environment, seeing camping as a way to offset these losses. When a New Zealand study asked parents at campsites to rank the importance of the family camping holiday on a scale of 1–5, where 5 was very important, 96% of respondents scored their response as 4 or 5 (Freeman and Kearns, 2015). Even when holidays may not be possible, days out to natural places also play an important role. These can be higher-order outings, such as to the seaside, the zoo or a nature reserve or more local and low-key visits, such as feeding the ducks, a walk to a local green space or kicking a ball in a neighbourhood school or sports field. Living in a city with natural features such as forests or beaches does not mean they are necessarily accessible, as some families may lack the means to go there. In such cases, organised trips can act as an important means of providing nature-related experiences where families are unable to do so (Figure 6.4).

Animals

There is little information available on how animals shape behaviour in families. In our research, we always ask about the animals that the child's family has, and invariably, we get enthusiastic descriptions of the animals, their names, their behaviours and sometimes their deaths. In her fascinating book *Beastly London: A History of Animals in the City*, Velten (2013) points to the myriad ways the lives of animals and people intersect. Some are historic, such as organ grinders' monkeys. However, others persist across time, such as feeding ducks in the local park or feeding pigeons in Trafalgar Square, London. The latter was a highlight for many children on trips to the capital until banned in 2003. At home, the keeping of pets, such as caged birds, dogs, cats, rabbits and others, shows no signs of abatement, possibly the reverse if the rise of chains of large warehouse-type pet supply stores is any indication (e.g., PetSmart and Petco, US; Pets at Home, UK; and Pet Barn, Australia). Pets are indeed a fundamental part of city children's lives:

> A recent survey by the American Veterinary Medical Association reported that 70% of all households with children younger than age 6 and 78% of all households with children older than age 6 had pets.
>
> (in Melson, 2003)

Pets can help children go out into nature and increase their independence. Children who have a dog are reported to be more likely to perceive that it is possible to walk to visit natural places, and parents are less fearful of likely harm coming to their children when out (Christian et al., 2012). This finding is supported by Melson (2003), who found children with a dog have higher physical activity levels. Yet, we know little about whether the experience of pet ownership makes children more likely to be interested in nature. Reason suggests that such an extrapolation is entirely reasonable, but evidence is lacking. Vining (2003) looked in some detail at caring for animals and its consequences, arguing that "What remains unanswered from these studies is whether caring about animals or the environment is a part of the restorative response" (p. 91). In other words, it is hard to prove that connection with pets has a direct nature-based connection response, although such a response seems likely. However, attachment to animals seems to be associated with higher scores on the New Ecological Paradigm scale, which is an internationally recognised indication of eco-centricity (in Vining, 2003). Further, Myers et. al. (2003) identified that for children, "perception of what animals need is a rich nexus for environmental education" and interaction with domestic animals can enhance this perception. Melson (2003) argues that direct interaction helps promote a 'bio-centric' approach to life for children. For children in some parts of the world, sharing the home or family compound with domestic animals, such as chickens or cattle (Figure 6.5), is the norm. Where animals such as monkeys, cattle, snakes and others are present in the neighbourhood, they become part of children's daily lives, and through their presence, children become more aware of natural processes (Figure 6.6).

In our *Natural Neighbourhoods for City Children* study, during interviews with children, pets, the antics of wildlife and visiting zoos are all frequently mentioned

Figure 6.5 Calf resting outside a family home in Thiruvananthapuram, Kerala.

Figure 6.6 A child (on the right of the photo) shows his excitement on encountering monkeys at a local temple in Kathmandu.

elements of their lives. These activities are not necessarily 'natural' in the true sense of the word but are, as Vining states,

> activities that help us to gain a sense of ourselves as natural entities, subject to natural forces. If intimate association with animals is in fact an attempt to reconnect with our natural world, then it may be possible to heal the human-nature split and approach the world in a spirit of cooperation and conservation.
> (2003, p. 90)

Thus, while evidence is lacking, it seems likely that animal ownership can provide meaningful ways for children to engage with natural processes, leading to enhanced empathy and understanding of other species and a better understanding of natural life cycles and interdependencies (Figure 6.7).

Figure 6.7 Pet ownership provides opportunities for enhancing potential nature connection and for walking around the neighbourhood and its open spaces.

The home

Here we look specifically at the home environment of the child. As cities change, become larger and harder to negotiate, have more traffic and are more compact with more high-rise buildings, the character of the home and its environs becomes extremely important. Chapter 5 looked at how the built environment has changed and how those changes over time create different types of living environments with varying opportunities for nature connection. The growing impact of these changes was highlighted in a study of children's play in Amsterdam. The study shows that in the post-war years, outside play was the norm and supported by the Amsterdam Urban Planning Department. Between 1947 and 1970, they created some 700 public play spaces (Verstrate and Karsten, 2011). This action was in part an attempt to encourage children off the streets. Indeed, such was the concern about children's street play that in 1942, a decree was issued by the Amsterdam Traffic Committee prohibiting play in many places outside the designated play areas. It was an early sign of growing concerns for children as traffic volumes rose in cities. At that time, not only was the house not seen as an appropriate play space but family size, overcrowding and lack of space made it unsuitable for play, especially energetic play. The home was where parents, usually mothers, exercised control. Families played board games or engaged in other 'calm and dirt-free' pastimes only occasionally (usually on a Sunday); thus, there were not many possibilities inside for active play (Karsten, 2005, p. 281). However, when interviewing parents today, Karsten found in some homes, children's outdoor activities, such as hut building or soccer, had become indoor activities (Karsten, 2005) with the home becoming a far more important part of children's lives.[3]

Karsten's findings regarding children's play historically are echoed in a study by Moore in 1986, where building interiors were the least mentioned favourite place for play (1986, p. 14). Play was an outdoor activity. The continuing decline in access to the outdoors for city children means children are now far more likely to spend increasing amounts of time in the home and in supervised situations. This reduces their potential to seek out independently natural spaces, such as parks or disused land. Reduced independent mobility for children increases their reliance on spaces closer to home; thus, the home environs become especially important in providing for children's access to space to play and to interact with nature. So, what then makes for a good home environment? An early, but still possibly the best summary we have on children's housing needs in terms of space, is provided by Cooper Marcus and Sarkissian (1986):

1 Children need safe, uninhibited outdoor play for their physiological and mental health.
2 Parents need to be able to allow their children outside without constant, close supervision.
3 The environment around children's homes needs to be safe from traffic, pollution and unnecessary physical and social hazards.

4 Children should be able to experience the pleasures of finding bugs, picking leaves, smelling flowers, collecting things and so on without their parents or the management harassing them. Through such contact with nature, they may develop, among other things, an understanding of basic ecological principles.
5 Children need easy, casual access to other children without a formal invitation to play.
6 Children need places in the communal environment that are undeniably their territories where they can expect to find other children.
7 Children need to create private spaces for themselves (for example, tree houses, forts or clubhouses) on wild or unmaintained ground away from public view.
8 Children need to be able to move around their home neighbourhoods safely and to take little trips farther and farther from home to gain a sense of independence.

While only point four makes specific reference to nature, the list identifies key prerequisites for making nature contact possible. The first is that children can access the outdoors. The second is the time spent outdoors is significantly influenced by access to other children. Outdoor play is an innately social activity, so the more children who are present locally, the more likely the child is to spend more time outdoors. Our *Natural Neighbourhoods for City Children* study identified that social connections were strongly associated with outdoor play, and an earlier neighbourhood study found the most social neighbourhoods had the highest levels of outdoor play (Freeman, 2010). The work of Mikkelsen and Christensen (2009) showed a similar connection, stating, "Children's mobility has to be recognized as primarily social and that companionship pervades every aspect of the activity" (p. 37). For girls, outdoor play "took place predominantly in the garden or in the vicinity of the home" (p. 47).[4] Thus, homes that encourage companionship and have green space close by are most likely to enhance nature connection opportunities.

Inside the home

We have already mentioned serendipitous nature, where nature makes its own way into the home, but homes can also be made to embrace nature inside as well as outside. Plants inside the home are an easy way for children to interact with nature, although many indoor plant species are exotics. Through caring for plants, children learn to understand their life cycles; their different characters and properties; their need for water, light and food; and, if they are lucky enough to have a carnivorous plant, about predator-prey relationships. There are also a number of creatures that can be kept close to home in the garden, on the balcony or even indoors. There are numerous advice sheets and sources of information available on how to keep mini-beasts such as bugs, stick insects and worm farms. In New

Zealand, swan plants complete with a caterpillar are sold in garden shops and at school fairs; children watch the caterpillar – its chrysalis and its emergence into a monarch butterfly, which is then released. Homes can be welcoming of nature. Trees planted close to the home can allow children to look at them from inside or from their bedroom, watching the tree in the wind or the rain or the birds that perch on them.

Spaces around the home

Domestic gardens and spaces adjacent to the home are vitally important, especially for very young children. Growth in traffic has reduced children's safe access to streets and public spaces around the home and increased reliance on private spaces or gardens. These spaces vary hugely in their natural content. Too often, such spaces have become paved over or made into monoculture lawns, sometimes with isolated trees, designed to facilitate easy maintenance rather than be inviting green spaces. These are especially prevalent around poorer or social housing. The trend towards hard landscaping and increased building of structures such as decks, studios and infill developments; subdivision of gardens; and paving front gardens to create car parking spaces, all reduce children's ability to connect to nature and nature's ability to inhabit the city successfully.

Even where some residential estates have experienced urban renewal, there has too often been little attempt to make the environment more natural. These nature-depauperate environments around the home, whilst common, are unnecessary, and high-density housing, including low socio-economic housing, can be nature-rich. The housing in Figures 6.8a to 6.8f show two contrasting approaches to housing development. The Figures 6a, b and c are in the UK that have been the focus of urban renewal programmes in Manchester, Newcastle and Plymouth. In none of these has any attention been given to making the landscape more natural or providing green spaces. In contrast, the houses in Figure 6d, e and f are in Amsterdam show similar high-density housing but housing where natural planting and green space has been prioritised. As high-rise and higher-density housing becomes increasingly common, and indeed is the norm in many inner-city areas and in high-density populations, such as in parts of Asia, the ability for children to be able to access nature in the public spaces around their homes becomes vital. The Amsterdam photos show that high-density does not have to mean nature-deprived.

Private gardens

The potential to interact with nature in the home environment is essential for children's well-being and the well-being of urban nature. Significant in the home environment is the street and the space immediately adjacent to the home. Various studies have estimated the home range of children, depending on age, to be about 300 to 500 metres (Freeman and Tranter, 2011), but most activity tends to be concentrated around the home. For younger children, the closest

Figure 6.8 Comparative housing redevelopments. Top row: UK redeveloped low-income housing that pays little attention to green space contrasted with low-income housing in Amsterdam that does pay attention to green space provision.

spaces to home are the most important, i.e., the garden, the pavement and the street in front of the home. Where the dominant housing form is low-density with a private garden, children can access nature relatively easily. Traditional gardens, even on a small scale, can be immensely nature-rich with respect to

Figure 6.9 Australian suburban garden with swing, cubby house (den) and possum 'house' just visible on lower left branch of the tree.

both plant and animal species. Through their everyday activities, playing, passing through or engagement in household tasks, children can encounter nature in their garden. In climbing a tree, they will encounter the birds, animals and insects living there (Figure 6.9). In playing on the lawn, they encounter grass and plant species, and they will feel wind, rain, cold, heat and other natural elements. Increasingly, homeowners are undertaking activities to attract wildlife into their gardens. For example, bird feeding and building ponds are common, and in most cities, some wild species seek refuge in gardens, habituating to human activities.

Gardens that are good for nature and children are under threat from several quarters. The first is the increasing replacement of vegetation with hard landscaping, either for 'aesthetic' purposes or through conversion to provide extended living, as in decks or patios, as well as the main culprit, transformation into car parking spaces. This prioritisation of car parks above garden space can be seen in many new developments as indicated in Figures 6.10a and 6.10b, which shows housing in Auckland, New Zealand, and Plymouth, UK. Various estimates on the rate of transformation of front gardens to car parks are

a

b

Figure 6.10 Loss of garden space to cars, Auckland, New Zealand, and Plymouth, UK.

available. One study in Ealing, London, estimates that the average front garden is now 68% hard landscaped, i.e., paved/non-vegetated (Ealing LA21, 2005). Interestingly, the Ealing Front Gardens Project in its '43 Reasons not to Pave', makes no mention of children's loss of play space. Not only are front gardens being converted to car parks but also hard landscaping is becoming more prevalent as a gardening form.

While the Ealing study focused just on front gardens, a Leeds study (Perry and Nawaz, 2008), looking at flooding and water runoff, found that between 1971 and 2004, the total impervious cover increased by 12.6%. The area covered by paved garden surfaces increased by 138%, accounting for 75% of the total increase in impervious surfaces, with the remaining 25% due to an increase in the 'buildings' and 'other impervious' categories. Nearly a quarter (23%) of the increase was in back gardens. The main reasons for the loss of vegetated spaces, identified across a number of studies, include car parking, increased car use and a trend towards hard landscaping and low maintenance, easy-care gardens (Bhatti and Church, 2001; Hitchins, 2007; Freeman et al., 2012). The trend towards hard landscaping continues to escalate the loss of vegetated space around children's homes. One of the most detailed studies on the loss of garden space is an Australian study which reports that between 1984 and 2003, the average house floor area rose by 40% to an average of 227.6 m^2, while average garden/lot size reduced from 802 m^2 to 735 m^2 (Hall, 2008, 2010). New-build homes in Australia and in New Zealand now frequently have lot sizes of around 400m^2, about half the lot size for homes built in previous generations. The greatest loss experienced is in vegetated garden and play space, exactly those spaces where children can connect most easily with the natural world.

The final trend is densification, where additional homes are built on garden space through subdivision of lots or through new-build at higher densities. Higher building density in itself is not necessarily problematic, as good design can still make available vegetated spaces, usually on a communal basis. However, much new-build is of poor design, focused on individual homes, where the built footprint is designed to maximise the house size at the expense of garden space. Such homes are often constructed on small plots of land with a view to maximising number of homes without regard to green space provision. This type of densification is often a response to the need to curb urban sprawl and development on greenfield sites on the edge of cities. It is an admirable aim but often poorly executed, as shown in Figure 6.11a. The photo shows an Australian housing development where the size of the house is constant or increasing, but the plot size is decreasing, resulting in high-density, low-rise housing with minimum private and communal green space. While the principle of densification of the existing built environment in the city in order to protect peri-urban green space is a positive one, it requires good design of the infill and redeveloped sites and the prioritisation of both private and communal natural spaces (Figure 6.11b and 6.11c).

a

b

c

Figure 6.11 A new development in Australia, densely packed single-storey properties, leaving minimal garden space. In contrast 6.11b and 6.11c high-density housing has been achieved in Stockholm, Hammarby Sjöstad, but retains good communal green spaces.

Building more natural homes

It is possible for homes at most densities to reflect natural values, although admittedly far more difficult at the highest densities as in cities such as Hong Kong or Madrid. There is now growing interest in how the building itself can be used as a green space, as with green roofs, green walls or internal green courtyard spaces. Functional systems, such as storm-water drains and canals, can be designed as green spaces (Figure 6.12), and collection points for storm water can become neighbourhood ponds. Designing with respect to the existing environment is something that the Netherlands does well. In the Netherlands, water is a major element of the environment that can be planned with or against. In some instances, as in land reclamation, where it is necessary to keep out seawater from reclaimed land, planning 'against' is required. However, there are instances where the water becomes part of the housing landscape and is immediately adjacent to the house. Children in IJburg, a new suburb of Amsterdam, were observed using kayaks parked at their home to move around (Figure 6.13a). This integration of water into the living environment is common. This is especially the case in new island suburbs built on reclaimed land, such as IJburg, built on six artificial islands in IJ Lake and intended for occupation by around 45,000 residents. Also in IJburg is a network of cycleways, which, combined with easy access to waterways, enhances access to ecologically rich settings close to children's homes as part of their everyday living environment (Figure 6.13a-d).

Figure 6.12 Water channels double as ecological sites and amenity areas in a new development in Yppenburg, Den Hague, Netherlands.

a

b

Figure 6.13 Children in Ijburg, Amsterdam, Netherlands, can easily access natural spaces through kayaks, cycling and as part of their everyday play.

c

d

Figure 6.13 (Continued)

Housing design is important in creating nature-responsive homes. In the traditional suburban home, gardens can be ecologically important, and the trend towards native plantings and more natural, less 'tidy' planting and towards wildlife-friendly gardens is an important step in the right direction. The ability to manipulate the environment to create spaces, huts, dens and opportunities to build, climb, dig or throw stones is an essential element in connecting with nature and cannot be provided in a highly regulated garden. Children need messy spaces or spaces they can be untidy in and reorder, and messy spaces are where insects, birds and wildlife are more likely to live. Where housing is higher density, private green spaces may be limited; however, there are some excellent examples where high-density housing still provides home-based natural green space. In parts of Europe, where higher housing densities have been and continue to be the norm, communal spaces can provide for such needs. They also have the advantage of communality and a higher potential for social interaction that further encourages outdoor play. In countries where high-rise is becoming the dominant housing form, especially parts of Asia, China and Singapore, the challenges are even more acute, but similar general design principles will apply. Higher-density housing can therefore still provide potential for nature contact close to home. Table 6.1 includes guidance on good housing design:[5]

Table 6.1 Good housing design

Housing design	*Outdoor space design*	*Green space*
Courtyard type spaces for very young children	Safety and privacy, sense of enclosure from roads and wider city	Vegetation at different structural levels, ground planting, shrubs, bushes and mature trees
Homes to directly link to the communal green space, front access rather than rear	Private space adjacent to the home becoming communal space further away	Pathways that are environmentally enhancing with materials or plantings
Sitting and relaxing spaces to encourage diverse all age use	Range of space types, formal, such as lawns and pathways for active play, and informal for greater wildlife engagement	Plantings to attract wildlife and possible features such as bird feeders, ponds, nest/bat boxes
Private or semi-private vehicle-free green space	Natural forms including rocks, slopes, running and standing water	Natural materials, e.g., logs for sitting on and relaxed management to allow wildlife to flourish
Communality – housing design that indicates identity and belonging to the housing complex	A neighbourhood plan where local streets link to parks and open spaces	Use of plantings to create a sense of safe enclosure and boundaries for children
Shared ecological vision and understanding by residents	Ecological planning features such as green walls, roofs, closed loop canal/water systems	Some areas designated and prioritised for wildlife, e.g., ponds, meadow, dense bushes

The homes children live in vary markedly, depending on location, building form, wealth or its absence, cultural norms and a myriad of other factors. Yet, natural connections can be made across this variety of housing forms either within the house by allowing nature inside the home or through engaging with nature in the garden, street or communal open spaces. Some environments can be particularly challenging in this respect, as in high-rises, but as Singapore shows, even in these challenging environments, green space provision and encounters with nature are possible. For families in many western countries, and to some extent globally, the trend is towards less outdoor play and less independent mobility, and with this comes an increase in adult-supervised activities and more home-based activities. Passive indoor play becomes increasingly prevalent and with it the potential for reduced nature contact. Greener homes need to be developed in tandem with access to the outdoors and embracing of nature in a seamless indoor-outdoor flow. The photos from Amsterdam (Figure 6.8) show how such flow can be achieved through designing homes that encourage green spaces right up to the home. These homes are designed to look into the gardens, which are home to many creatures if children look hard enough. The snails in the final photo for this chapter (Figure 6.14) came from a garden and were brought into the kitchen for closer examination. These children were intrigued by the snails. They had not previously paid them much attention, but it only took a little encouragement for snail hunting to occur. Direct experiences with wildlife provide a vital basis on which to build lifelong nature connections, and the home and its surrounds play a critical role in this respect.

Figure 6.14 Looking at snails found in the garden.

Notes

1 Tolerance of geckos does vary by individuals and by culture and country.
2 A reference to the common holiday pursuit of catching crabs in beach rock pools.
3 The notion that children should only play in designated playgrounds with pre-formed play structures is being challenged. Streets are being increasingly treated as legitimate informal play spaces, and play in natural places is being strongly encouraged (see description of Jeugdland Maakland play area in Amsterdam, chapter 11).
4 While generally other children act as a pull for outdoor play, negative interactions can also inhibit this activity.
5 Some of these examples are based on ideas in (Cooper Marcus, 2012; Beatley, 2011; Coles and Millman, 2013).

References

Academy of Science. 2009. *Jane Goodall Interview Academy of Science*, August 20, 2009. www.achievement.org/autodoc/page/goo1int-5

American Experience. 2016. *American Experience: Theodore Roosevelt and the Environment*. www.pbs.org/wgbh/americanexperience/features/general-article/tr-environment/ accessed 2–10–2016.

Beatley, T. 2011. *Biophilic Cities: Integrating Nature into Urban Design and Planning*, Island Press, Washington.

Bhatti, M., and Church, A. 2001. Cultivating natures: Homes and gardens in late modernity. *Sociology* 35(2): 365–383.

Christian, H., Trapp, G., Lauritsen, C., Wright, K., and Giles-Corti, B. 2012. Understanding the relationship between dog ownership and children's physical activity and sedentary behaviour. *Pediatric Obesity* 8: 392–403.

Coles, R., and Millman, Z. (eds.). 2013. *Landscape, Well-Being and Environment*. Routledge, Abingdon, Oxon, UK.

Cooper Marcus, C. 2012. Planning for a silent minority: The needs of children for outdoor play, access to nature, and independent mobility. In T. Haas (Ed.), *Sustainable Urbanism and Beyond: Rethinking Cities for the Future*, Rizzoli, New York.

Cooper Marcus, C., and Sarkissian, W. 1986. *Housing as If People Mattered: Site Design Guidelines for Medium-Density Family Housing*. University of California Press, Berkeley.

Department of Conservation. 2006. *Review of Camping Opportunities in New Zealand: Report to the Minister of Conservation*, Department of Conservation, Wellington.

Ealing's LA21. 2005. *Ealing Front Gardens Project*. www.ealingfrontgardens.org.uk

Freeman, C. 2010. Children's neighbourhoods, social centres to 'terra incognita'. *Children's Geographies* 8(2): 157–176.

Freeman, C., Dickinson, K.J.M., Porter, S., and van Heezik, Y. 2012. 'My garden is an expression of me': Exploring householders' relationships with their gardens. *Journal of Environmental Psychology* 32: 135–143.

Freeman, C., and Kearns, R. 2015. Childhoods under canvas: Campgrounds as spaces of resistance to protective parenting. *Childhood* 22(1): 101–120.

Freeman, C., and Tranter, P. 2011. *Children and the Urban Environment: Changing Worlds*, Earthscan, London.

Hall, T. 2008. Where have all the gardens gone? *Australian Planner* 45(1): 30–37.

Hall, T. 2010. *The Life and Death of the Australian Backyard*, CSIRO Publishing, Collingwood, Australia.

Hitchins, R. 2007. How awkward encounters could influence the future form of many gardens. *Transactions of the Institute of British Geographers* 32: 363e376.
Karsten, L. 2005. It all used to be better? Different generations on continuity and change in urban children's daily use of space. *Children's Geographies* 3(3): 275–290.
Lusher, A. 2014. Interview with David Attenborough. *Daily Mail Australia*. www.dailymail.co.uk/femail/article-2569361/Sir-David-Attenboroughs-childhood-memories.html
Melson, G.F. 2003. Child development and the human-companion animal bond. *American Behavioral Scientist* 47: 31–39.
Mikkelsen, M.R., and Christensen, P. 2009. Is children's independent mobility really independent? A study of children's mobility combining ethnography and GPS/mobile phone technologies. *Mobilities* 4(1): 37–58.
Moore, R.C. 1986. *Childhood's Domain: Play and Place in Child Development*, Croom Helm, London.
Myers, O.E., Jr., Saunders, C.D., and Garrett, E. 2003. What do children think animals need? Aesthetic and psycho-social conceptions. *Environmental Education Research* 9(3): 305–325.
Packham, A. 2015. Family Holidays Are the Source of Happy Memories, and the Benefits Don't Stop After You Come Home 23/06/2015 12:13. *The Huffington Post UK*, 7 December 2015. www.huffingtonpost.co.uk/2015/06/23/family-holidays-happy-memories_n_7643456.html
Perry, T., and Nawaz, R. 2008. An investigation into the extent and impacts of hard surfacing of domestic gardens in an area of Leeds, United Kingdom. *Landscape and Urban Planning* 86: 1–13.
Velten, H. 2013. *Beastly London: A History of Animals in the City*. Reaktion Books, London.
Verstrate, L., and Karsten, L. 2011. The creation of play spaces in twentieth-century Amsterdam: From an intervention of Civic actors to a public policy. *Landscape Research* 36(1): 85–109.
Vining, J. 2003. The connection to other animals and caring for nature. *Human Ecology Review* 10(2): 87–99.

7 Children, nature and health

The benefits of nature

Why health?

One of the primary catalysts for the developing interest in children's connection to nature has been the rising concern for children's well-being or, more specifically, children's perceived lack of well-being, where lack of connection to nature is seen as a direct contributory factor. Health dis-benefits have been identified as an increase in a whole range of health disorders, most commonly physical disorders, such as obesity, and mental health disorders, such as depression, attention deficit hyperactivity disorder (ADHD), and inattentiveness. Louv coined the term 'nature deficit disorder' to describe this set of health disorders, which he defines as follows:

> The human costs of alienation from nature, among them: diminished use of the senses, attention difficulties, and higher rates of physical and emotional illnesses. The disorder can be detected in individuals, families and communities . . . deficit is only one side of the coin. The other is natural abundance. By weighing the consequences of the disorder, we can also become more aware of how blessed our children can be biologically, cognitively, and spiritually – through positive connection to nature.
>
> (Louv, 2008, p. 36)

Regardless of whether we accept Louv's notion of nature deficit disorder, Louv and others have drawn increasing attention to the relationship between children, nature and well-being. It is this debate and the relationship nature has with children's well-being that is the focus of this chapter. We look at how this relationship has been conceptualised by key thinkers (especially the notion of biophilia), by eco-psychologists and others. First, we set the conceptual framework, and then we move to examine the specific benefits and dis-benefits of nature connection and how these relate to urban living and urban form.

Conceptualising the nature–health relationship: biophobia to biophilia

For much of history, connection with nature has been seen as an inevitable part of human existence. To live apart from nature, even in urban areas, was difficult and if achieved was the preserve of very few people living in a few of the larger urban conglomerations. Further, even in cities, animals such as cats, horses, rats and sheep coming to market (Velten, 2013) would have been part of everyday city life. It would have been hard historically for children in any living environment to be totally separated from nature. The need for nature and the form of the nature relationship have become the foci of increasing attention, in part possibly due to the need to better understand this relationship as living environments, particularly urban environments, undergo changes. Eric Fromm (in Wilson, 1993) coined the term *biophilia* for "the passionate love of life and all that is alive". Progressing this notion further, Wilson developed an idea that has become known as the biophilia hypothesis, namely, "the innately emotional affiliation of human beings to other living organisms . . ." (Wilson, 1993, p. 31). Kellert further dissects the biophilia notion and asserts that the "human need for nature is linked not just to the material exploitation of the environment but also to the influence of the natural world on our emotional, cognitive, aesthetic and even spiritual development . . ." (1993, p. 42). The biophilia notion, therefore, powerfully asserts that much of the human search for a coherent and fulfilling existence is intimately dependent on our relationship with nature. Using this biophilia notion, Kellert then goes on to develop a typology of nine biophilic values. These values were derived for a generic adult population but can be usefully applied to children's nature relationships. We have adapted Kellert's table to include a column in which the explanations children provided in our *Natural Neighbourhoods for City Children* study are included. These inclusions relate specifically to responses they gave to questions on 'what nature is' and 'why they like and dislike it' and are placed next to the nine values of Kellert with which they are most closely associated.

Through using the biophilia hypothesis, researchers have given weight to the need for people to affiliate with nature if they are to experience well-being. It is an idea that has also found common cause with the public. The biophilia hypothesis is, however, not universal in its acceptance. Joye and de Block critique the scientific value of the biophilia hypothesis and argue that serious inquiry is needed to offset what they see as the uncritical acceptance of the biophilia hypothesis in popular and academic literature (2011, p. 190). They challenge the evolutionary justification inherent in the biophilia hypothesis, but importantly for this book, they still concur with the notion that people indeed affectively relate to life-like elements and processes (p. 208). Several studies have found complementary results citing in excess of two-thirds of respondents valuing nature and believing in its right to exist (van den Born et al., 2001). The notion of people's affinity to nature finds commonality across populations and places; it is this affinity to and the benefits associated with nature that we focus on in our discussion of children's biophilia.

At the other end of the continuum to biophilia is biophobia. As people became more able to manipulate their living environments, the role of nature has become less evident. In more extreme cases, this leads to the presence of biophobia, when feelings range from 'discomfort in natural places' to the culturally acquired urge to affiliate with technology, human artefacts and solely with human interests as regards the natural world (Orr, 1993, p. 416). The last of these is of particular relevance to the child-nature relationship as in the dominance of technology over other experiences (Figure 7.1a and 7.1b). Digital and related devices are understood to

Table 7.1 A typology of biophilia values after Kellert (1993) with examples given by children in our *Natural Neighbourhoods for City Children* study

Term	*Definition*	*Function*	Natural Neighbourhoods for City Children *study – children's responses to the question: why do you like/dislike nature?*
Utilitarian	Practical and material exploitation of nature	Physical sustenance/ security	Likes eating fruit from gardens* You can make food out of it (*exploitation*) Like eating huhu grubs, bugs, like playing around, biking, building huts, bike jumps (*sustenance and practical exploitation of natural materials and landforms*) There might be an apple tree there that has apples (*sustenance*)
Naturalistic	Satisfaction from direct experience/ contact with nature	Curiosity, outdoor skills, mental/physical development	Because you get to explore new things, play in the grass, hear the birds (*curiosity*) Climbing trees (*physical development*) Exploring, always something new to see (*outdoor skills*) You can climb trees and make tree houses and play in the grass at parks (*satisfaction, direct experience*)
Ecologistic-scientific	Systematic study of structure, function and relationship in nature	Knowledge, understanding, observational skills	Like seeing new things, Dad studies horticulture and is teaching me (*study*) Lots of bees that come to pollinate flowers (*observation and knowledge*) Like seeing different types of plants, animals, nature (*observation*)
Aesthetic	Physical appeal and beauty of nature	Inspiration, harmony, peace security	You get to see butterflies and pretty insects (*physical appeal*) Air is fresh, it's really green (*inspiration and beauty*) Peaceful, quiet and colourful (*harmony, beauty, inspiration*)
Symbolic	Use of nature for metaphorical expression, language, expressive thought	Communication mental development	Smells good, normally there's a story to every plant (*communication*) There's a lot of spikey bits, trees grow over, so light is dappled (*expressive language and thought*) Life force that lives all around us (*mental development, expressive thought*)

(*Continued*)

Table 7.1 (Continued)

Term	*Definition*	*Function*	Natural Neighbourhoods for City Children *study – children's responses to the question: why do you like/dislike nature?*
Humanistic	Strong affection, emotional attachment, 'love' of nature	Group bonding, sharing, cooperation, companionship	It's always peaceful and you can hear birds chirping and it's really calm (*emotional attachment*) Like to listen to the trees (*emotional attachment*) Calm, can lie down and look up and see lots of things, can walk around and look at nature (*emotional attachment*) It's good for playing hide and seek and games with friends (*companionship*)
Moralistic	Strong affinity, spiritual reverence, ethical concern for nature	Order and meaning in life, kinship and affiliational ties	Because Grandad has absolutely amazing crib, going to go camping there (*kinship*) Grandma has nice trees and flowers (*kinship*) Fresh air, feels connected to the Earth (*affinity to earth*)
Dominionistic	Mastery, physical control and dominance of nature	Mechanical skills, physical prowess, ability to subdue	Because all the nature in the backyard has been destroyed so they could build two new houses behind the house (*dominance of nature, ability to subdue*) Usually sees lots of buildings that tear apart the nature, and he doesn't really like that (*dominance of nature, ability to subdue*) There's big trees to climb (*mechanical skills*)
Negativistic	Fear, aversion, alienation from nature	Security, protection, safety	Nature can stink. Like mud and old birds' nests (*aversion, alienation*) Venomous animals that sneak up (*fear, safety*) Too many bugs (*aversion*) Creepy crawlies (*aversion*) Getting lost (*safety*)

* Some are the children's own words, but some are taken from summarised conversations with the researcher; table adapted from Kellert, 1993

a

b

Figure 7.1a At Jurong Bird Park in Singapore, for some, the emphasis seems to be on the technological media rather than on the wildlife present; the man in the photo 7.1b watched the whole bird show through his iPad screen.

be a major factor in the reduction of children's nature connection (see Chapter 5). Further, the technology relationship is seen as having its own health impacts as well as its association with reducing the health benefits associated with outdoor play and connection to nature.

The biophilia concept has resonated with the eco-psychologists who tie in the intimate connection with nature to personal and planetary well-being. Roszak encapsulated this in the following image:

> The Earth hurts, and we hurt with it. If we can accept the imagery of a Mother Earth, we might say that the planet's umbilical cord links us at the root to the unconscious mind.
>
> (1992, p. 308)

Eco-psychology, now a field of psychology in its own right, focuses on the 'circle of care' in nature relationships and increasingly focuses not so much on individual connection but planetary needs and connections. The biophilia concept has also found common cause with advocates of a return-to-nature approach to childhood. Childhood is seen as the time when innate biophilic tendencies are strongest, and their awakening enables children to become imbued with a lifelong desire for association with nature. Over time, they become adults who are themselves nature-centred. This relationship is a two-way process associated with children's own growing concerns about and developing ability to act on behalf of nature and the wider planetary environment. Simultaneously, a biophilic relationship develops that enables children to realise the health benefits that can accrue from this relationship. The ways that this relationship develops is complex, and how the tie develops is poorly understood and, according to Wilson, is 'crying out for scientific enquiry' (Wilson, 1993, p. 40). It is not known, for example, how much and what type of experiences are necessary beyond obvious ones: for example, is direct and continuing contact better than indirect and occasional contact in developing biophilia?

It is not known either how biophilia develops in urban environments and through modern lifestyles or whether these necessarily act as barriers against biophilic growth. In addition, while we can assume at an individual level the importance of small personal nature interactions, according to Soule, while "Our little acts of biophilia – buying bird seed, caring for our pets, nurturing our gardens – sustain us emotionally . . . what we need now are big, selfless, and costly acts of biophilia to protect nature" (Soule, 1993. p. 453). These big acts, however, are the responsibility of adults. Yet too often, there are statements to the effect that children today need to understand nature better so they can tackle the environmental problems of the future. While to some extent, adults offloading this task onto children is inevitable and children's nature connection is seen as a preparation for taking on this responsibility, this expectation is unfair and irresponsible on the part of today's adults.

Biophilia for children requires nurturing through contact with natural features and environments. However, their experience of these will be mediated through a cultural milieu, populated by individuals, and takes place in cultural settings whose influence on the developing biophilia of children and young people has been little explored or understood by biophilia exponents. Just as nature itself is diverse, so too are children and the places and cultures they live in, meaning there is no one biophilic pathway, and for children living in urban areas similarly, there is diversity not universality of experience. Within this diversity though are some commonly recognised health benefits associated with nature connection for children that relate to children's physical, cognitive, emotional and social well-being.

The health benefits of nature

The relationship between green space, contact with nature and health has been of growing interest to researchers across several disciplines. An early paper by Frumkin (2001) looked at findings from studies on people in a whole range of settings, including offices, prisons, hospitals, dental surgeries and psychiatric facilities, that all suggested that contact with the natural world, even indirectly (e.g., view from a prison cell window), resulted in health benefits. As a consequence, he concluded, "As we learn more about the health benefits of contact with the natural world, we need to apply this knowledge in ways that directly enhance the health of the public" (2001, p. 239). In their review of the evidence for interactions with nature providing benefits for people, Keniger et al. (2013) laid out typologies of the settings (indoor, urban, city fringe, production landscape, wilderness), interactions and potential benefits of human experiences with nature (Table 7.2). Nature interaction has been linked to improved cognitive and affective functioning and mental health; walks in natural areas or proximity to green space are associated with decreased anxiety, stress and rumination, increased working memory performance and improved mental health and vitality (Bratman et al., 2015a; van den Berg et al., 2016). However, most studies to date are correlational, and little is known of the causal relationships and mechanisms that deliver the benefits (Gascon et al., 2015). As Lee and Maheswaran observe (2010), although there is plenty of assertion that green spaces are good for health "establishing the causal relationship has been difficult" (2010, p. 212).

One idea proposed to explain restorative effects on cognitive function tested by psychologists Berman et al. (2008) is that the many stimuli provided by a natural environment, such as a sunset, require our attention in a modest manner. This allows our directed-attention abilities a chance to recover, so that after exposure to nature, we are better able to perform tasks that depend on these capabilities. In this way, nature has

Table 7.2 Typologies of interactions with nature and health benefits (after Keniger et al., 2013)

Health benefits of interaction	*Types of interactions*
Psychological well-being	Direct (e.g., getting wet from rain, hearing birdsong)
Cognitive	Indirect (e.g., view from a window, wildlife films and images)
Physiological	Incidental (as a by-product when doing something else, e.g., seeing flowers while walking to school)
Social	Intentional (hiking, gardening, swimming in a stream)
Spiritual	
Tangible	

a restorative effect. A different approach was taken by environmental neuroscientists, who used a controlled experiment to show that nature experience reduces rumination activity, that is, repetitive thoughts focused on negative aspects of the self, and this was accompanied by reduced neural activity in the part of the brain linked to risk of mental illness (Bratman et al., 2015b). These experimental studies provide a deeper understanding of the significance of nature contact for human well-being.

Two large-scale studies, both from the UK, that provide more evidence of this beneficial nature relationship are as follows. The first is based on 1,138,481 responses from 21,947 UK respondents. Using a smartphone app and global positioning system (GPS) position data, people's levels of happiness were recorded along with their environment. It was found that "study participants are significantly and substantially happier outdoors in all green or natural habitat types than they are in urban environments" (Mackerron and Mourato, 2013, p. 992). The second study used mortality data from a study population of 40,813,236, including 366,348 deaths, and assessed this against the quantity of green space in their residential area. The data came from the UK's generalised land use database. Their findings revealed that even for low-income populations, there is lower mortality in populations who live in the greenest areas, thus green space helps lower mortality rates. They went on to conclude that "interventions in the physical environment are highly effective at affecting health and health behaviours" (Mitchell and Popham, 2008, p. 1659).

Other studies focus on more specific aspects of the nature–health relationship, and here, we use just a few to illustrate the wealth of research being undertaken in this respect and across a wide range of geographic settings. In Finland, Korpela and Yle'n (2007) found favourite natural settings were associated with reduced stress. Greater emotional benefits were reaped by those with health complaints. In a fascinating study with elderly residents of a nursing home in China involved in growing indoor plants, Tse found "significant improvements in life satisfaction, social network and a significant decrease in perception of loneliness for older people in the experimental group after the eight week indoor gardening programme" (2010, p. 949). Also addressing older adults, Bossen (2010) looked at how nature-based stimuli can help dementia patients. What is interesting about these nature and health research examples is that there is no comparable research that identifies nature as being bad for health, except when nature itself is compromised, as in pollution or in relation to extreme nature events, such as floods or earthquakes. The overwhelming evidence is, therefore, of nature as a health benefit.

While these studies identified specific health benefits, there are some commonly recognised benefits for people in general. When these are typologised, the spread of benefits accruing from nature contact is revealed. Although these are benefits for adults, it is reasonable to assume they are also are indicative of the type of benefits available to children. We now focus on how the benefits identified in the following typology specifically relate to children.

Children's health and nature connection

Contact with nature contributes to the health and well-being of children in many ways, beginning at birth – physical health (general health, neonatal weight and survival, respiratory diseases and allergies, physical fitness), cognitive function and self-control, psychological well-being, affiliation and imaginative play and affiliation with other species and the natural world (reviewed in Chawla, 2015). Studies in Germany and the US have identified an association between behavioural problems in children, particularly hyperactivity/inattention problems, and poor access to urban green spaces (Faber Taylor et al., 2001; Markevych et al., 2014). Among inner-city children living in high-rise buildings, the naturalness of the view from their homes was associated with enhanced self-discipline in the case of girls, who were more likely (than boys) to spend time playing in and around their homes (Faber Taylor et al., 2002). In Spain, a study of 2,593 7- to 10-year-old children in Barcelona showed improved cognitive development over a 12-month period in children exposed to greater greenness around their homes and schools and during commuting (Dadvand et al., 2015). Wells's study with 17 children from low-income families in the US showed children who relocated to homes in greener environments demonstrated improvements in attentional capacity and cognitive functioning (Wells, 2000).

Despite the recognised benefits of natural spaces on health, research over the last decade points towards a decline in both children's preferences for natural places and access to these places (Korpela et al., 2002; Loukaitou-Sideris, 2003; Min and Lee, 2006). These studies show most children as preferring formal play and sport settings, places where suitable play objects are available and where the site itself is conducive to the kinds of activities they wish to engage in. Spaces close to home have become increasingly important to many children. Poorer areas are usually associated with reduced access to green space (Richardson et al., 2010); the environment itself is usually of poorer quality, and poor neighbourhoods have more crime and traffic, more hazardous play spaces and poorer services (in Evans, 2004). Because children in poorer neighbourhoods are less likely to participate in organised activities, they are more likely than children from wealthier neighbourhoods to play outdoors and within their own neighbourhoods. Their selection of places to play reflects safety threats as much as the positive characteristics the various places afford. Where there are safety concerns relating to outdoor places, school gardens and domestic gardens can be especially important sites for outdoor play.

Green schoolyards provide a space where children can escape stress, develop competencies and social connections and deepen their connection to nature, enhancing their resilience throughout childhood and adolescence (Chawla et al., 2014; Wake and Birdsall, 2016). In particular, vegetation volume in schoolyards has been positively associated with perceived restorativeness (Bagot et al., 2015). The smaller land area commonly associated with poorer housing usually contains less garden space (Mathieu et al., 2007). This is especially so for children living in denser accommodation, such as apartments (Whitzman et al., 2010). It can be difficult for children to acquire the health benefits associated with green space when they live in high-density environments that are lacking in outdoor green spaces and when safety concerns limit their access to the green spaces that are present.

Emotional benefits are implicit in children's association with nature. Early childhood educators use words such *awe*, *wonder*, *joy* and *inner peace* when describing children's relationships with nature. These attributes are generally associated with beautiful settings and in relation to contact with natural objects, such as seashells and rocks (Schein, 2014). Focus-group interviews with 68 children in the US revealed an intellectual and emotional appreciation of nature, with children expressing their relationship with nature as a friendship or love (Kalvaitis and Monhardt, 2015). A number of authors who have examined the relationship between happiness and nature conclude that not only does nature produce positive moods, but it also promotes environmentally appropriate behaviours (Nisbet et al., 2011; Mackerron and Mourato, 2013; Zelenski and Nisbet, 2014). These studies focused on adults, and while there are no directly comparable studies on children's happiness, it is reasonable to assume a similarly positive relationship. For children, two important correlates of happiness are social interactions and play, both important components of well-being. In our *Natural Neighbourhoods for City Children* study, when asked why they liked nature, happiness was clearly a by-product of the experience, as indicated in the quotes in Table 7.3. However, only one child made direct reference to happiness, saying that people "are always happy and not fighting each other [when in nature]". The children's statements in Table 7.3 reveal the full range of health benefits identified in the typology of well-being. Children, even in relatively short statements, can allude to multiple and complementary well-being benefits from being in nature.

Table 7.3 Application of the typology of health benefits linked to children's reported enjoyment of nature, given in response to the question: why they liked being in nature during our *Natural Neighbourhoods for City Children* study.

Child's response to "Why do you like being in nature?"	*Psychological well-being*	*Cognitive*	*Physiological*	*Social*	*Spiritual*	*Tangible*
Fun exploring, something new to see	☐		☐			
Because you get to explore new things, play in the grass, hear the birds			☐			☐
Fun things to do when you're bored, peaceful, comforting	☐					
Feel free, nobody there, just self	☐				☐	
There aren't many people around, and it's peaceful, can hang out with friends				☐	☐	
It's fun, when there's big trees you can climb there and hang out with friends and can find new places and build stuff there	☐		☐	☐		☐

Child's response to "Why do you like being in nature?"	*Psychological well-being*	*Cognitive*	*Physiological*	*Social*	*Spiritual*	*Tangible*
Like it when it's windy, because likes the sound of the trees and the birds	☐					☐
I like it because you can always hear the birds sing	☐					☐
Likes eating huhu grubs, bugs, likes playing around, biking, building huts, bike jumps						☐
Quiet, and you can find interesting things		☐			☐	
It's nice and sunny, you can get exercise, there's plenty of room to run about if you're feeling energetic			☐			☐
There aren't many people around, and it's peaceful, can hang out with friends				☐	☐	
It's refreshing	☐				☐	
Can explore and find new things, like plants		☐				☐
Hearing the birds, it's nicer air than in the city			☐			☐
It's cool	☐					
Peaceful, beautiful, and good hiding places	☐				☐	☐
Peaceful, safe	☐				☐	
Fun, and there's lots of stuff to do	☐					☐
Because being stuck inside is like all stuffy, and being outside you get fresh oxygen and it's more fun outside	☐		☐			
When it's windy . . . sound of the trees and the birds			☐			☐
Likes seeing new things, Dad studies horticulture and is teaching her		☐				☐
Sometimes sees something she's never seen		☐				

Outdoor play

There has been a significant decline in children's outdoor play and independent mobility in many countries, especially in parts of the developed world. Substantial reductions in independent mobility across three generations have been identified in a New Zealand study, which shows that children today are not permitted the same unrestricted licence to roam that their parents and grandparents experienced (Bhosale et al., 2015). Parents' understandings as to why these changes had occurred touch on many ways that changing social policies have affected community and family life (Witten et al., 2013). A recent study highlights the decline in independent mobility in a very stark and visual way (Woolley and Griffin, 2015). The study was a small one carried out with two families, comprising grandparents, parents and children who had all lived their lives in the same Sheffield suburb in the north of England. In the one family, the grandparents, when they were children, could travel up to 3 km from home (measured as the crow flies) without asking permission. The second family grandparent could also go to 3 km but needed permission for places over 1.5 km away. The grandchild (age 6) of the first grandparent was only allowed to go to a friend some 25 m away with permission. In the second family, the grandchild (age 10) could go 115 m away to a friend's with permission. Neither child could visit natural places without an adult.

This growing level of constraint is echoed in an earlier 1995 multi-generation study in Leeds, a city north of Sheffield. Parents were interviewed and asked about their childhoods, where they went and what they did, and then about the places their own children could go. Here, similar restrictions were noted. The parents could essentially free-roam anywhere within and adjacent to the suburb, a range of up to five miles being usual from about age 9, but further trips were also mentioned. The only place mentioned by participants as being restricted was swimming in the quarry, as their parents believed it was dangerous. Interestingly, the wide roaming range was present amongst both boys and girls. When the study was undertaken in 1995, only one child, whose parents were also interviewed, had full freedom of the suburb. Most could go to the local park, but a few could access only their own street. One parent whose own children aged 8 and 9 could only play in the front garden compared this with her own childhood:

> I think my children miss out on the sort of adventure play which was my particular favourite, fantasy adventure in old buildings, building fires and cooking outside, dens and staying out all day. I was also more independent – for example, going to the cinema alone to meet friends, or catching a bus to Hunslet swimming baths on my own.
>
> (Freeman, 1995, p. 272)

The favourite places mentioned for the parents' generation included a number of natural and semi-natural places: Oulton quarry, the pit heaps, the Goit (a water channel), a canal, streams, a local hill, parks, woods, a local mansion and grounds and various local lakes and ponds. Activities such as insect collecting, blackberry picking, hide-and-seek, fishing and finding tadpoles, tree climbing, cycling, egg collecting, picking flowers, conkers (horse chestnuts), sledging and swimming were common. Most of these activities were not possible for the parents' own children, as they were confined to much smaller ranges close to home and within the confines of the neighbourhood. The reduced range can be seen in Figure 7.2, which indicates where children living in the suburb were observed playing during site visits in the school summer holidays.

The limited independent mobility of children today has been identified in a number of studies, together with an associated decline in outdoor play, particularly play beyond the family home. The decline is evident across several countries, including Australia, the Netherlands, Scandinavia, the UK and the United States (Fyhri et al., 2011; Karsten and van Vliet, 2006, Foster et al., 2014; Kyttä et al., 2015). An international comparative study of children's independent mobility found children in Australia had far less independent mobility than children in Japan, Tanzania and South Africa (Malone and Rudner, 2011). The variation in mobility they found "is

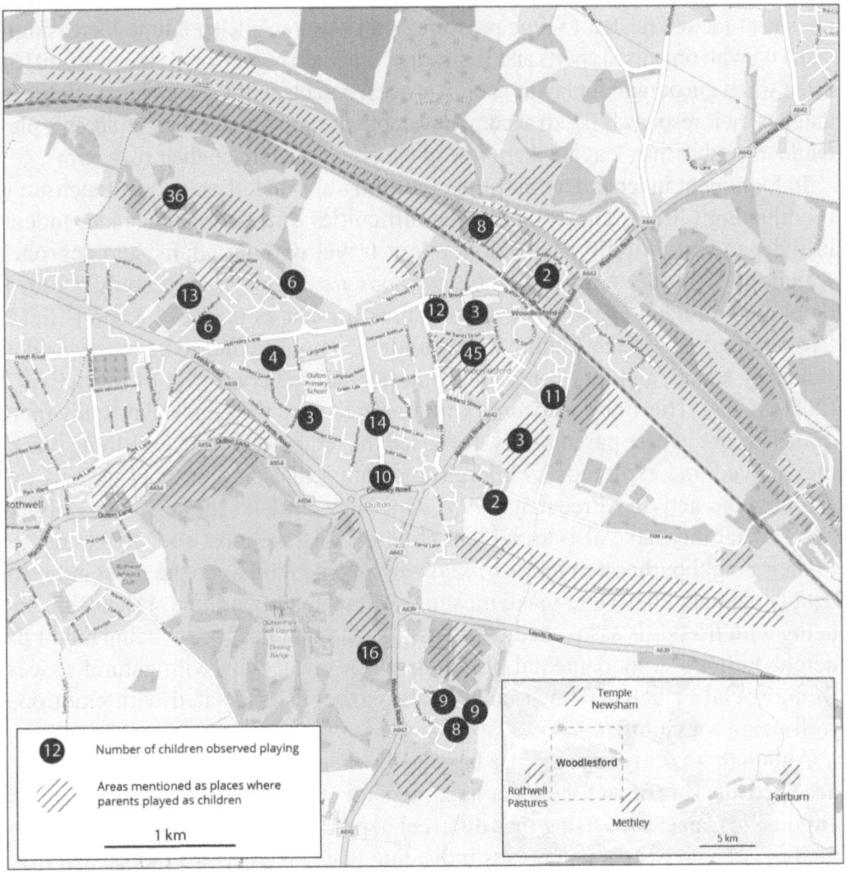

Figure 7.2 The Leeds study showing the contrast between the wider area used by parents (hatched area) and the smaller, more localised play spaces used by children in 1995 (black circles).

influenced by the social, cultural and physical characteristics of a particular community context. It is at a community level that children engage in mobility, and it is at this local social scale that negotiations of independence and mobility are made between children and adults" (Malone and Rudner, 2011, p. 254). Perception of danger is one of the common reasons for limiting children's independent mobility. It can be argued that perception of danger is related to social perception more than 'actual danger'. Although there is plenty evidence of increased traffic densities, the evidence for a significantly more dangerous environment in terms of stranger danger for children is yet unconvincing. Statistically, the greatest danger of harm and abduction for children is from family members or people they know, rather than from strangers (Newiss and Collie, 2015). Kyttä et al. (2015) suggest that an important cultural characteristic promoting high childhood independent mobility in Finland is the degree of shared responsibility or trust among neighbours and even strangers, with stranger-danger being among the least-often selected reasons for picking a child up from school in their study. Other significant barriers to children's mobility to engage in nearby nature are social factors related to time pressure, with children's leisure time increasingly taken up with organised sports and leisure activities and homework (Skar et al., 2016). Even when encouraging play in natural spaces, parents in this Norwegian study felt that it is their responsibility to accompany, motivate and support their children's play outdoors, rather than leaving them free to experience nature on their own terms.

In looking at independent mobility, the study by Mikkelsen and Christensen is illuminating (2009). They point to the difficulties of defining the 'independent' in independent mobility: Is it alone? Is it travel without adults (the approach we adopted in our *Natural Neighbourhoods for City Children* study)? Can it be play in an adult space but without adults present as in use of a sports field? Less ambiguous is their finding that "children's mobility has to be recognised as primarily social, and that companionship pervades every aspect of the activity" (2009, p. 37). This finding accords with findings from the Leeds play study mentioned before, where parents only ever used the term 'we' when describing places they played and explored as children, indicating that outdoor play was always a communal activity (Freeman, 1995). Our New Zealand studies on neighbourhood sociability have also found that children who are most socially connected to their neighbourhood in terms of knowing people tend to be the most independent (Freeman, 2010) and make most use of the neighbourhood. Friendship and being with friends is associated with positive parental licence for being out in the neighbourhood, a fundamental prerequisite for connecting with natural spaces. Being social or being with friends is also strongly associated with emotional well-being for children.

Although it is easy to recognise happiness when we see it, there is no generally accepted measure of well-being or happiness for children. However, working with children in Canada and using four different happiness measures, Holder and Coleman found a consistency in results indicating that engagement in social activities and high-quality friendships are related to higher levels of happiness. However, spending time alone was associated with lower levels of happiness (2009). A comparative study in India found similar results in that "children who were more social and active, and less shy were happier" (Holder et al. 2012, p. 261). In addition to the often quoted stranger-danger fears and traffic increase as factors inhibiting contact with and play outdoors, societal changes where it is more difficult for children to socialise can impact on their ability to access the outdoors. Factors such as changing societal ways of interacting (remotely through technology), increasing numbers of children spending time away from their neighbourhood by attending more distant schools, spending more time on educational activities and sports or cultural or other supervised activities and time in formal child-care facilities all leave less time for informal socialising and outdoor play (Freeman, 2010).

What do these societal changes mean for nature connection and health? Are children less able to connect with natural places though exploratory or adventure play? The opportunities available to previous generations in Leeds and Sheffield are not available for children with extremely limited independent mobility or for children who lack access to friends with whom they can play outdoors. The impacts seem to be greater for girls, who can have lower rates of independent mobility and outdoor play (Larson et al., 2011). Our *Natural Neighbourhoods for City Children* study also shows that the trend of declining independence and play has a negative impact particularly on children's ability to access natural places. In our study, we asked children to place 40 dots on an aerial photograph of their home and neighbourhood, placing the most dots in the outdoor places they spent the most time. The pattern that emerged was striking in that children spend the most time in their gardens, then on sports fields, and the least time in the most natural spaces, namely woodland and spaces with non-woodland natural habitats (Figure 7.3a and 7.3b). In some ways, this is reflective of the fact that the children had limited independent mobility and thus limited home ranges, so made most use of spaces close to home, invariably the garden. However, even where there were natural spaces close by, these were not used by the children as much as expected given their proximity and size. Children can thus find it difficult to acquire the health benefits associated with nature spaces.

a b

Figure 7.3a This park with its limited biodiversity was well used by children in the *Natural Neighbourhoods for City Children* study, as it was close to their homes, whereas more biodiverse sites, such as a nearby bush reserve, 7.3b, were little used.

Familiar animals

The focus so far has been on experience of place, but in talking to children in all of our research projects, one theme that comes up repeatedly is the role of animals. Animals have an innate fascination for children. In our studies with children, none of which have been animal-focused, it has become very clear that animals form an important part of their nature connection. Yet animals – in the sense of animals present in daily life, such as pets, and local wildlife, such as garden birds or everyday species, including bees or spiders, have been largely ignored in terms of their role in nature connection. A recent study in Japan that asked 1,030 urban residents to assess the likeability of 29 wild animals and their willingness to co-exist with these animals in their neighbourhoods showed that childhood encounters with nature, particularly in the form of collecting insects and plants, was the most important factor in determining positive attitudes towards wild animals (Hosaka et al., 2017). Childhood interactions with plants and animals are therefore vital in shaping attitudes towards nature in later life.

While the health benefits of animals for adults have been highlighted many times, these benefits invariably relate to companion animals consistently, cats and dogs, with an emphasis on assistance animals. Frumkin (2001) quotes a number of studies that quantify animal-related benefits, in terms of, for example, reduced number of doctors' visits and lower levels of stress. Maller et al. (2005) examined the evidence around the therapeutic benefits of animal interaction. They concluded, "[R]esearch demonstrates with certainty . . . [that] . . . There are some known beneficial physiological effects that occur when humans encounter, observe or otherwise positively interact with animals, plants, landscapes or wilderness" (2005, p. 50). While there is a large body of research on the therapeutic benefits of animals, especially companion animals for adults, there is very limited research on children, even though animal-assisted interventions are becoming increasingly used in the field of medicine (Brazier, 2014). Animals have been used to assist children with emotional and physical disorders ranging from autism and hearing impairments to cerebral palsy (Esposito et al., 2011). An Australian study found children in dog-owning families were less likely to be overweight than comparative non-dog-owning families (Timperio et al., 2008), and dog walking by children has been identified as an effective way of increasing independent mobility as well as physical activity (Christian et al., 2014). Certainly, pets have been commonly associated with enhanced levels of emotional empathy and do provide emotional support for children, especially when under stress (Faber Taylor and Kuo, 2006). While these studies do illustrate that people relate at an affective level to animals, it is not clear whether this response is biophilic, i.e., indicative of an innate relationship with living organisms. Therapeutic effects might be attributable to how the animals look or feel, rather than their presence alone (Joye et al., 2011).

Children are naturally inclined to gravitate to animals as is very clear at zoos, farms or through their interest in pets, such as dogs (Figure 7.4a), cats, rabbits or other common domesticated animals, and the younger the animal the greater the interest tends to be (Figure 7.4b). There has been little research done on

a

b

Figure 7.4 Animal interactions: (a) Walking the dog with Grandad; (b) an urban farm, Temple Newsam, Leeds, UK; (c) Tom and his favourite chicken; (d) Leo, Tiffany and Caz the assistance dog.

c 

Photo by Mandy Tocher

d 

Figure 7.4 (Continued)

Photo by David Wiltshire

children's interactive relationships with wild animals. The following examples relate to domesticated animals and are suggestive of the likelihood that connection with wild animals would have positive benefits for children, but there is a need for research to be able to support this suggestion. The boy in Figure 7.4c holding the chicken is particularly fond of the chickens living on his large edge city property. He has made a particular friend of a 'disabled' chicken. Of all the chickens, this one that has a defective beak is his favourite and the one to which he pays particular attention. Animals seem to be important vehicles for supporting children's development of empathy and a caring ethic. It is a relationship that can also go two ways, as animals can also respond positively to children who are more vulnerable. Leo, the older of the two children in Figure 7.4d, has multiple disabilities, and his parents notice his particular empathy for animals (Figure 7.4d). In 2015, the family, with Leo, then 6, and his sister, Tiffany, 4, who both have Down syndrome and related health issues, added Caz, a dog from Assistance Dogs New Zealand, to their family. Their mother noticed, "Caz had a calming influence on the children and so they wanted to be near her" (Knight, 2015). Further, Caz makes it easier for the family to access outdoor places and go for walks, as each child has his or her own lead for Caz, which encourages the children to walk and helps keep them safe. While it is to be assumed that contact with wild animals will provide similar positive emotional benefits, this is to date largely conjecture, as there have, as far as we can determine, been no studies on the health benefits of interaction with wild creatures. So far, we have focused on the health benefits that relate to interaction with nature. However, it is also necessary to recognise that for a large number of children, there is a converse relationship, where their environmental interactions may not be beneficial to their health but rather are characterised by poor levels of physical and personal safety.

Challenging environments

The environment can severely affect children's health, especially where that environment itself is compromised, for example, by pollution. According to the WHO (2016 ref), an estimated 12.6 million deaths each year are attributable to unhealthy environments, with the numbers of children being disproportionately represented and poor water quality being a major contributor to poor health outcomes. Diarrhoeal disease is the third leading cause of death among children under five. The WHO (2016) estimates that more than 340,000 children under five die annually from diarrhoeal diseases due to poor sanitation, poor hygiene or unsafe drinking water – that is almost 1,000 per day. For children living in some of the world's most difficult urban living conditions, access to green space and natural environments is marginal to non-existent, or its presence is a source of dis-benefits rather than benefits for people's health. For example, Dharavi (Figure 7.5), one of the world's largest slums and located in central Mumbai, has a population estimated at between 300,000 and 1,000,000 and has a large child population. The slum became known internationally through the 2008 film *Slumdog Millionaire*, which featured

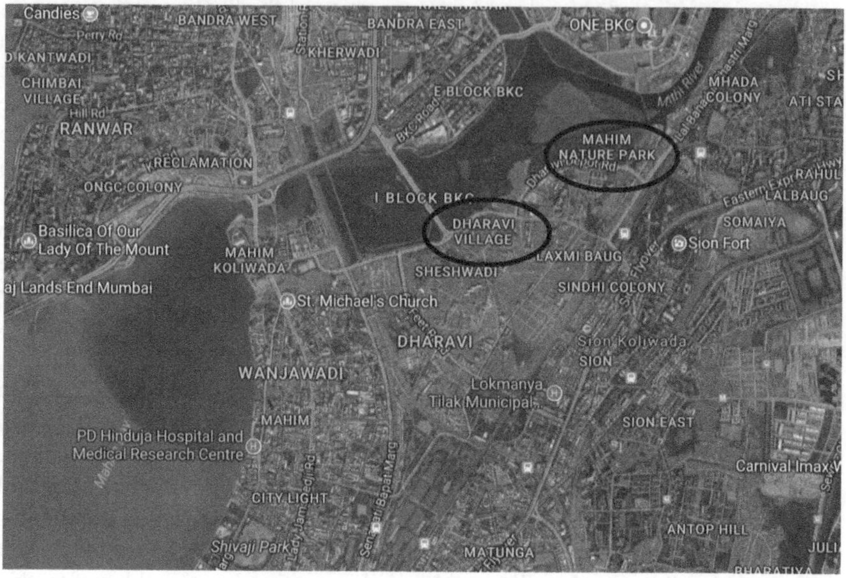

Figure 7.5 Dharavi slum Mumbai showing the location of the nearby Mithi River and in the top right the Mahim Nature Park (Google Maps).

children from the slum. Children growing up there will have very limited access to green space, and the local Mahim Creek that has been created from land previously used as a dump, rather than being an ecological asset, is a source of contagious diseases; it also contains heavy toxic metals and industrial effluent (Singare and Ferns, 2014). However, even in these most challenging circumstances, green space is available locally and improvement is happening. Mahim Nature Park adjoins the slum and was created on land that until the 1970s was used as a garbage dump. The Mithi River, again heavily polluted, borders the settlement but has nonetheless wetland and woodland vegetation. Redevelopment proposals for Dharavi seek to introduce green features, especially along roads and public gathering spaces. However, although green spaces are locally available and areas of more natural environments may be present, this does not mean they will be perceived as safe and be used by children.

The United Nations Educational, Scientific and Cultural Organisation's (UNESCO's) Growing Up in Cities Project interviewed children in Soweto, a poor, predominantly black African settlement in Johannesburg, South Africa's largest city. Children spoke of avoiding the veld (open grassland areas) and bushy places, which were seen as unsafe, yet good use was made of a far less natural small grassed lawn area by a busy road and garage (Swart Kruger, 2002). The presence of people enhanced the area's feeling of safety. A more

Figure 7.6 (a) Suva Point, Suva, Fiji, well-used by local children, contrasts markedly with (b) a beach frequented by tourists.

recent study of 1,004 12-year-old participants living in Cape Town reinforced the UNESCO findings for poorer children: Children from middle-income communities have opportunities and access to nature, evincing a close attachment to nature (Adams and Savahl, 2015), which acts as a buffer against life stressors and helps in developing resilience proficiencies (Wells and Evans, 2003). Children from low-income communities, however, experienced nature as the dangerous other, as natural spaces are characteristically unsafe and dangerous (Adams et al., 2016).

In a study of urbanisation and its impact on children in Fiji (Freeman et al., 2015), the children disliked the wetland areas they had to traverse to get to school, as they could be muddy and difficult to cross during Fiji's frequent heavy rain periods. The local beach in the capital, Suva Point (Figure 7.6a), was well used by city children, but a visit there reveals that the beach is muddy and the shoreline heavily strewn with garbage, a very different experience from that provided by the white-sand tourist beaches (Figure 7.6b). In contrast to Dhavravi though, vegetation is present in the informal settlements in Suva, indicating that poverty is not automatically associated with a lack of more natural elements (Figure 7.7). For children in poor and environmentally challenged environments, nature can be a source of positive connection but can also exacerbate the problems children living in these areas already face and impact negatively on their health and well-being.

Figure 7.7 A well-vegetated informal settlement in Fiji showing that poor housing is not necessarily deprived in respect of vegetation.

Accessing nature's benefits

There is considerable evidence that contact with nature and time spent in natural places is good for people's emotional, cognitive and physical health and well-being. The range of health benefits is broad. Especially important for children are the emotional benefits associated with happiness, fun and the sense of adventure, social well-being from being with friends, enjoyment from nature, curiosity, peace and spiritual well-being. There are also physical benefits associated with freedom in open spaces and challenging features, such as trees and hills to climb and streams to play or swim in (Figure 7.8a and 7.8b). In order for children to interact with nature, they have to be able to spend time in natural environments. Concerns about safety and risk have been a major factor inhibiting children's ability to connect to nature and experience associated health benefits. For children living in environmentally challenging environments, natural areas and features, such as rivers, can themselves be health reducing, especially where these features themselves have been compromised by urban waste and despoliation. On the positive side, chiefly for children living in challenging urban environments, natural spaces can provide precious relief from some of the stresses that occur elsewhere in their lives. Natural environments are therapeutic environments and environments that encourage developmental progress, but the most important health benefit of all is that they are environments that nearly always make children happy.

Figure 7.8a Many cities have rivers that provide excellent opportunities for children to have fun with nature, Suva, Fiji.

Figure 7.8b A very different type of river built as part of the South Bank complex, Brisbane, Australia, and designed to mimic a natural waterway is always popular with the children.

References

Adams, S., and Savahl, S. 2015. Children's perceptions of the natural environment: A South African perspective. *Children's Geographies* 13(2): 196–211.

Adams, S., Savahl, S., and Casas, F. 2016. The relationship between children's perceptions of the natural environment and their subjective well-being. *Children's Geographies*. DOI: 10.1080/14733285.2016.1157571

Bagot, K.L., Allen, F.C.L., and Toukhsati, S. 2015. Perceived restorativeness of children's school playground environment: Nature, playground features and play period experiences. *Journal of Environmental Psychology* 41: 1–9.

Berman, M.G., Jonides, J., and Kaplan, S. 2008. The cognitive benefits of interacting with nature. *Psychological Science* 19(12): 1207. DOI: 10.1111/j.1467–9280.2008.02225.

Bhosale, J., Duncan, S., Schofield, G., Page, A., and Cooper, A. 2015. A pilot study exploring the measurement of intergenerational differences in independent mobility. *Journal of Transport & Health* 2: 522–528.

Bossen, A. 2010. The importance of getting back to nature for people with dementia. *BC Journal of Gerontological Nursing* 36(2).

Bratman, G.N., Daily, G.C., Levy, B.J., and Gross, J.J. 2015a. The benefits of nature experience: Improved affect and cognition. *Landscape and Urban Planning* 138: 41–50.

Bratman, G.N., Hamilton, P., Hahn, K.S., Daily, G.C., and Gross, J.J. 2015b. Nature experience reduces rumination and subgenual prefrontal cortex activation. *PNAS* 112(28): 8567–8572.

Brazier, A. 2014. Creature comfort? Animal assistance in therapeutic work with children and young people. *Clinical Child Psychology and Psychiatry* 19(2): 165–168.

Chawla, L. 2015. Benefits of nature contact for children. *Journal of Planning Literature*: 1–20. DOI: 10.1177/0885412215595441

Chawla, L., Keena, K., Pevec, I., and Stanley, E. 2014. Green schoolyards as havens from stress and resource for resilience in childhood and adolescence. *Health & Place* 28: 1–13.

Christian, H., Trapp, G., Villanueva, K., Zubrick, S.R., Koekemoer, R., and Giles-Corti, B. 2014. Dog walking is associated with more outdoor play and independent mobility for children. *Preventive Medicine* 67: 259–263.

Dadvand, P., Niewenhuijsen, M.J., Esnaola, M., Forns, J., Basagaña, X., Alvere-Pedrerol, M., Rivas, I., López-Vicente, M., De Castro Pascual, M., Su, J., Jerrett, M., Querol, X., and Sunyer, J. 2015. Green spaces and cognitive development in primary schoolchildren. *PNAS* 112: 7937–7942.

Esposito, L., McCune, S., Griffin, J.A., and Maholmes, V. 2011. Directions in human-animal interaction research: Child development, health, and therapeutic interventions. *Child Development Perspectives* 5(3): 205–211.

Evans, G.W. 2004. The environment of childhood poverty. *American Psychologist* 59(2): 77–92.

Faber Taylor, A., and Kuo, F.E. 2006. Is contact with nature important for healthy child development? State of the evidence. In C. Spencer and M. Blades (Eds.), *Children and Their Environment*, Cambridge University Press, Cambridge.

Faber Taylor, A., Kuo, F.E., and Sullivan, W.C. 2001. Coping with ADD: The surprising connection to green play settings. *Environment and Behaviour* 33(1): 54–77.

Faber Taylor, A., Kuo, F.E., and Sullivan, W.C. 2002. Views of nature and self-discipline: Evidence from inner-city children. *Journal of Environmental Psychology* 22: 49–63.

Foster, S., Villanueva, K., Wood, L., Christiana, H., and Giles-Corti, B. 2014. The impact of parents' fear of strangers and perceptions of informal social control on children's independent mobility. *Health & Place* 26: 60–68.

Freeman, C. 1995. The changing nature of children's environmental experience: The shrinking realm of outdoor play. *The International Journal of Environmental Education and* Information 14(3), July–September: 259–281.

Freeman, C. 2010. Children's neighbourhoods, social centres to 'terra incognita'. *Children's Geographies* 8(2): 157–176.

Freeman, C., Lingam, G.I., and Burnett, G. 2015. Children's changing urban lives: A comparative New Zealand-Pacific perspective. *Journal of Urban Design* 20(4): 507–525. DOI: 10.1080/13574809.2015.1044508

Frumkin, H. 2001. Beyond toxicity: Human health and the natural environment. *American Journal of Preventive Medicine* 20(3): 234–240.

Fyhri, A., Hjorthol, R., Mackett, R.L., Fotel, T.N., and Kyttä, M. 2011. Children's active travel and independent mobility in four countries: Development, social contributing trends and measures. *Transport Policy* 18(5): 703–710.

Gascon, M., Triguero-Mas, M., Martinez, D., Dadvand, P., Forns, J., Plasència, A., and Nieuwenhuijsen, M.J. 2015. Mental health benefits of long-term exposure to residential green and blue spaces: As systematic review. *International Journal of Environmental Research and Public Health* 12: 4354–4379.

Holder, M.D., and Coleman, B. 2009. The contribution of social relationships to children's happiness. *Journal of Happiness Studies* 10: 329–349.

Holder, M.D., Coleman, B., and Singh, K. 2012. Temperament and happiness in children in India. *Journal of Happiness Studies* 13: 261–274.

Hosaka, T., Sugimoto, K., and Numata, S. 2017. Childhood experience of nature influences the willingness to coexist with biodiversity in cities. *Palgrave Communications*. DOI: 10.1057/palcomms.2017.71

Joye, Y., and de Block, A. 2011. 'Nature and I are two': A critical examination of the biophilia hypothesis. *Environmental Values* 20: 189–215.

Kalvaitis, D., and Monhardt, R. 2015. Children voice biophilia: The phenomenology of being in love with nature. *Journal of Sustainability Education* 9, March.

Karsten, L., and van Vliet, W. 2006. Children in the city: Reclaiming the street. *Children, Youth and Environments* 16(1): 151–167.

Kellert, S.R. 1993. The biological basis for human values of nature. In S.R. Kellert and E.O. Wilson (Eds.), *The Biophilia Hypothesis*, Island Press, Washington, DC.

Keniger, L.E., Gaston, K.J., Irvine, K.N., and Fuller, R.A. 2013. What are the benefits of interacting with nature? *International Journal of Environmental Research and Public Health* 10: 913–935. DOI: 10.3390/ijerph10030913

Knight, L. www.stuff.co.nz/manawatu-standard/news/70166020/Assistance-dog-to-improve-lives-of-Down-syndrome-kids-Leo-and-Tiffany accessed 13–7–2015.

Korpela, K.M., Kyttä, M., and Hartig, T. 2002. Restorative experience, self-regulation, and children's place preferences. *Journal of Environmental Psychology* 22: 387–398.

Korpela, K.M., and Yle´n, M. 2007. Perceived health is associated with visiting natural favourite places in the vicinity. *Health & Place* 13: 138–151.

Kyttä, M., Hirvonen, J., Rudner, J., Prirjola, I., and Laatikainen, T. 2015. The last free-range children? Children's independent mobility in Finland in the 1990 and 2010s. *Journal of Transport Geography* 47: 1–12.

Larson, L.R., Green, G.T., and Cordell, H.K. 2011. Children's time outdoors: Results and implications of the National Kids Survey. *Journal of Park and Recreation Administration* 29(2): 1–20.

Lee, A.C., and Maheswaran, R.J. 2010. The health benefits of urban green spaces: A review of the evidence. *Public Health* 33: 212–222. DOI: 10.1093/pubmed/fdq068. Epub.

Loukaitou-Sideris, A. 2003. Children's common grounds: A study of inter-group relations among children in public settings. *Journal of the American Planning Association* 69(2): 130–143.

Louv, R. 2008. *Last Child in the Woods: Saving Our Children from Nature-Deficit Disorder*. Algonquin Books, Chapel Hill, NC.

MacKerron, G., and Mourato, S. 2013. Happiness is greater in natural environments. *Global Environmental Change* 23: 992–1000.

Maller, C., Townsend, M., Pryor, A., Brown, P., and St. Leger, L. 2005. Healthy nature healthy people: 'Contact with nature' as an upstream health promotion intervention for populations. *Health Promotion International* 21(1): 45–54.

Malone, K., and Rudner, J. 2011. Global perspectives on children's independent mobility: A socio-cultural comparison and theoretical discussion of children's lives in four countries in Asia and Africa. *Global Studies of Childhood* 1(3). Doi.org/10.2304/gsch.2011.1.3.243

Markevych, I., Tiesler, C.M.T., Fuertes, E., Romanos, M., Dadvand, P., Nieuwenhuisjen, M.J., Berdel, D., Koletzko, S., and Heinrich, J. 2014. Access to urban green spaces and behavioural problems in children: Results from the GINIplus and LISAplus studies. *Environmental International* 71: 29–35.

Mathieu, R., Freeman, C., and Aryal, J. 2007. Mapping private gardens in urban areas using object-oriented techniques and very high-resolution satellite imagery. *Landscape and Urban Planning* 81(3): 179–192.

Mikkelsen, M.G., and Christensen, P. 2009. Is children's independent mobility really independent? A study of children's mobility combining ethnography and GPS/mobile phone technologies. *Mobilities* 4(1): 37–58. DOI: 10.1080/17450100802657954

Min, B., and Lee, J. 2006. Children's neighbourhood place as a psychological and behavioral domain. *Journal of Environmental Psychology* 26: 51–71.

Mitchell, R., and Popham, F. 2008. Effect of exposure to natural environment on health inequalities: An observational population study. *Lancet* 372(9650): 1655–1660.

Nisbet, E.K., Zelenski, J.M., Murphy, S.A. 2011. Happiness is in our nature: Exploring nature relatedness as a contributor to subjective well-being. *Journal of Happiness* 12: 303–322.

Newiss, G., and Collie, C. 2015. *Police-Recorded Child Abduction and Kidnapping England, Wales and Northern Ireland 2012/13 to 2013/14 Parents and Abducted Children Together*, Parents and Abducted Children Together, London.

Orr, D.W. 1993. Love it or lose it: the coming biophilia revolution. In S.R. Kellert and E.O. Wilson (Eds.), *The Biophilia Hypothesis*, Island Press, Washington, DC.

Richardson, E., Pearce, J., Mitchell, R., Day, P., and Kingham, S. 2010. The association between green space and cause-specific mortality in urban New Zealand: An ecological analysis of green space utility. *BMC Public Health* 10(1): 240. http://doi.org/10.1186/1471-2458-10-240

Roszak, T. 1992. *The Voice of the Earth*, Simon and Schuster, New York.

Schein, D. 2014. Nature's role in children's spiritual development. *Children, Youth and Environments* 24(2): 78–101.

Singare, P.U., And Ferns, S.E.L. 2014. Study of toxic heavy metals in Mahim Creek of Mumbai. *International Letters of Chemistry, Physics and Astronomy Online* 36: 98–106. DOI: 10.18052/www.scipress.com/ILCPA.36.98

Skar, M., Wold, C., Gundersen, V., and O'Brien, L. 2016. Why do children not play in nearby nature? Results from a Norwegian study. *Journal of Adventure and Outdoor Learning* 16(3): 239–255. DOI: 10.1080/14729679.2016.1140587

Soule, M.E. 1993. Biophilia unanswered questions. In S.R. Kellert and E.O. Wilson (Eds.), *The Biophilia Hypothesis*, Island Press, Washington, DC.

Swart Kruger, J., and Chawla, L. 2002. 'We know something someone doesn't know': Children speak out on local conditions in Johannesburg. *Environment & Urbanization* 14(2): 85–96.

Taylor, A.F., Kuo, F.E., and Sullivan, C. 2002. Views of nature and self-discipline: Evidence from inner city children. *Journal of Environmental Psychology* 22: 49–63.

Timperio, A., Salmon, J., Chu, B., and Andrianopoulos, N. 2008. Is dog ownership or dog walking associated with weight status in children and their parents? *Health Promotion Journal of Australia* 19: 60–63.

Tse, M.M. 2010. Therapeutic effects of an indoor gardening programme for older people living in nursing homes. *Journal of Clinical Nursing* 19: 949–958.

van den Berg, M., van Poppel, M., van Kamp, I., Andrusaityte, S., Balseviciene, B., Cirach, M., Danileviciute, A., Ellis, N., Hurst, G., Masterson, D., Smith, G., Triguero-Mas, M., Uzdanaviciute, I., de Wit, P., van Mechelen, W., Gidlow, C., Grazuleviciene, R., Nieuwenhuijsen, M.J., Kruize, H., and Maas, J. 2016. Visiting green space is associated with mental health and vitality: A cross-sectional study in four European cities. *Health Place* 38: 8–15.

Van den Born, R.J.G., Lenders, R.H.J., de Groot, T., and Hijsman, E. 2001. The new biophilia: An exploration of visions of nature in Western countries. *Environmental Conservation* 28(1): 65–75.

Velten, H. 2013. *Beastly London: A History of Animals in the City*, ReaKtion Books, London.

Wake, S.J., and Birdsall, S. 2016. Can school gardens deepen children's connection to nature? In K. Nairn, P. Kraftl, and T. Skelton (Eds.), *Space, Place, and Environment: Geographies of Children and Young People* 3: 89–114. DOI: 10.1007/978-981-287-044-5_1

Wells, N. 2000. At home with nature: Effects of 'greenness' on children's cognitive functioning. *Environment and Behaviour* 32(6): 775–795.

Wells, N.M., and Evans, G.W. 2003. Nearby nature: A buffer of life stress among rural children. *Environment and Behavior* 35: 311–330.

Whitzman, C., Romero, V., Duncan, M., Curtis, C., Tranter, P., and Burke, M. 2010. Links between children's independent mobility, active transport, physical activity, and obesity. In E. Waters, B. Swinburn, R. Uauy, and J. Seidell (Eds.), *Preventing Childhood Obesity: Evidence, Policy, and Practice*, Wiley-Blackwell, Hoboken, NJ. [ISBN 978-1-4051-5889-3 [4]

Wilson, E.O. 1993. Biophilia and the conservation ethic. In S.R. Kellert and E.O. Wilson (Eds.), *The Biophilia Hypothesis*, Island Press, Washington, DC.

Witten, K., Kearns, R., Carroll, P., Asiasiga, L., and Tava'e, N. 2013. New Zealand parents' understandings of the intergenerational decline in children's independent outdoor play and active travel. *Children's Geographies* 11(2): 215–229.

Woolley, H.E., and Griffin, E. 2015. Decreasing experiences of home range, outdoor spaces, activities and companions: Changes across three generations in Sheffield in north England. *Children's Geographies* 13(6): 677–691. DOI: 10.1080/14733285.2014.952186

World Health Organisation. www.who.int/mediacentre/news/releases/2016/deaths-attributable-to-unhealthy-environments/en accessed March 2016.

World Health Organisation. www.who.int/water_sanitation_health/monitoring/jmp-2015-key-facts/en/

Zelenski, J.M., and Nisbet, E.K. 2014. Happiness and feeling connected: The distinct role of nature relatedness. *Environment and Behavior* 46(1): 3–23.

8 Culture and nature, different experiences, different ways of interacting

Socio-cultural contexts of nature

Children's relationship with nature and their ability to access nature can be dependent not just on the presence or absence of nature in the places they live but also on socio-cultural practices. Aspects such as the national psyche and the place of nature in that psyche, cultural practices or individual factors, such as gender, age or ability, help to shape the child-nature relationship. Understanding these influences is fundamental. If the focus on enhancing nature connection and experience is on providing and enhancing opportunities located within the physical environment, but the factors determining that relationship are located in the social, cultural or personal environment, then a mismatch will occur, regardless of any biodiversity initiatives. The western arcadian/romantic lens has dominated much of the approach to viewing the child-nature relationship (see Figure 1.3) and is rooted in a particular western cultural context. Its current renaissance through the back-to-nature movement strongly reflects this particular lens. The often unquestioning acquiescence to the western romantic approach is increasingly being contested and indeed should be, if meaningful child-nature relationships that respect the myriad of geographic and socio-cultural contexts in which children's lives are lived are to be supported. In this chapter, we look at how cultural diversity influences children's ways of interacting with the natural world and their experiences.

In her book in which she 'reconfigures the natures of childhood', Affrica Taylor refers to what she calls 'enacted nature cultures'. She points to the diversity of children and their varied 'multinatural' worlds. She argues there is no one natural world that children encounter but rather complex and varied nature worlds. Cities and societies are becoming increasingly diverse and plural, with children belonging to and moving between multiple societal groups and cultures. The differences present in these societal and cultural groups influence children's experiences of nature, and some (different) experiences will be favoured over others, e.g., going fishing or collecting seafood may be prioritised over a walk in a forest or nature reserve. The need to recognise the ethnic, cultural background and other socially determined factors that influence children's relationships with nature can be seen in Auckland, New Zealand's largest city. Auckland's population is ethnically diverse and becoming increasingly diverse, with approximately 790,000 NZ

Europeans/Pākehā;[1] 143,000 Māori; 195,000 Pacific Islanders; 307,000 Asians and 25,000 from the Middle East, Latin America or Africa (Auckland Council, 2016). Immigration from other countries continues to boost the multi-ethnic nature of the city, which inevitably influences directly how children view and experience nature. In Auckland, therefore, there is and will increasingly be no one way for nature connection to occur, and if differences in culture, ethnicities and places of origin are ignored, the nature connection story will be incomplete.

In looking at nature, Selin states, "All people everywhere transform nature" (2003, p. xix). Nowhere is this change more apparent than in cities. Processes of change are shaped by the knowledge that is transmitted through social learning:

> Most knowledge is culturally transmitted and shaped by environmental and social forces which vary from place to place. In part, what people know is constrained by local ecology, although what is uniquely human is the capacity for acquired biological knowledge to diffuse independently of what can be experienced in local habitats.
>
> (Ellen, 2003, p. 47)

It is vital, therefore, to explore cultural and social underpinnings and the different cultural lenses children use. These differences have not only shaped the urban environment but also continue to determine who uses that environment and how.

Children see the world through different lenses, not only through the lens of childhood but through a lens initially determined by cultural, socio-economic, ethnic, religious and other factors present in the family. As the child's world broadens, the family lens is added onto or replaced by a societal one reflecting wider societal norms. Difficulties can arise where the family lens and the wider social lens are at variance with each other. This matters for nature connection, as the family, the society and the character of the individual need to be, if not in harmony, at least closely aligned for nature interaction to be supported. Societal views can be central determinants of a child's nature experience, and one consequence of a discrepancy between family and societal norms can be withdrawal from nature interaction. On the positive side, national views on nature and associated cultural practices can embed and encourage nature relationships. The various national green strategies[2] in Singapore have resulted in the city's strong network of parks and green spaces, including greening of transport routes, thus making natural features an integral part of the urban environment. But it is the Singaporean cultural, social and economic[3] fabric that will guide the frequency and type of interaction children have with these natural spaces. Within countries, differences can emerge, often needing to be explained by reference to socio-cultural factors. A study of children's neighbourhoods in Suva, Fiji's capital city, found Indo-Fijian[4] children, especially girls, were least likely to play outside their home, whereas indigenous Fijian boys were most likely to play outdoors. Indigenous Fijian children were also more likely than Indo-Fijian children to swim in the rivers or in the sea (Freeman et al., 2015). For indigenous Fijian boys, working on the family plantation/garden and for girls, helping in the home are tasks that curtail the time available for outdoor play and activities, such as river swimming. In Fiji then, ethnicity, cultural roles and gender all affect connection with and opportunities to experience natural spaces.

In this chapter, we begin by looking at different viewpoints, focusing on three main themes; the first is relationships with nature at the national level, usually as an embodiment of a sense of nationhood, but also reflective of religious and cultural beliefs. Second is an exploration of how children's sense of identity and belonging, associated cultural and social practices and levels of ability/disability affect relationships with nature. For some children, as with indigenous children, the challenges of urban life can be particularly stark. The third is nature as 'avoidance' or nature as 'threat'. This chapter identifies some of the myriad ways children can experience nature based on their different backgrounds and explores how this difference can help explain both the difficulties children may confront in their nature relationships but also how to support diverse ways of being in nature. Figure 8.1 summarises some of the key differentiating and homogenising factors in children's nature relationships. The diversity of children's experiences is immense so this chapter can unfortunately only identify a few of the many influential factors governing children's socio-cultural lives.

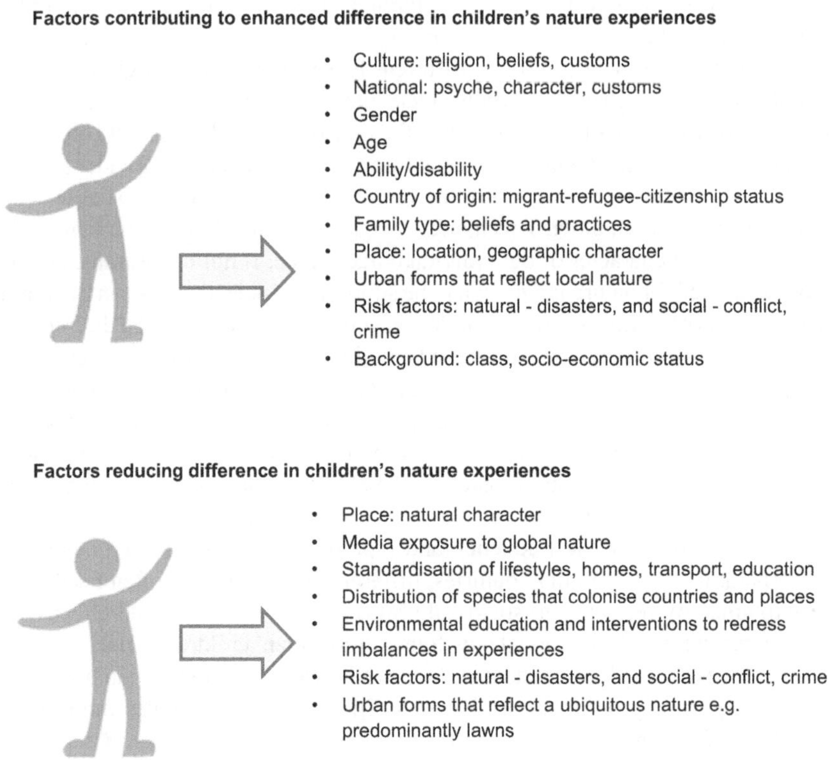

Figure 8.1 Difference and nature.

Different viewpoints

The process of globalisation is homogenising urban nature as people, plants and animals make their way into the urban environment. Therefore, perhaps it is apposite to expect that children will have encounters with increasingly similar nature. Several species have populated cities around the world; the house sparrow, the rock pigeon and the Indian mynah are now found in cities across many continents and islands. In a study of cities in 36 countries on all continents, four bird species were found to occur in more than 80% of cities: rock pigeon (51 cities), house sparrow (48 cities), European starling (44 cities), and barn swallows (43 cities). For plants, 11 occur in more than 90% of cities, with the most common, the annual meadow grass (*Poa annua*) occurring in 96% of the cities and in all biogeographic realms (Aronson et al., 2014, p. 4). As described in Chapter 3, the nature present in cities is highly modified, with an increasing presence of globally widespread species. Thus, even where there are strongly rooted national or ethnic-based nature traditions, philosophies and relationships, these may not translate well into an urban context where nature has been modified from its native state. However, on the positive side, while many species are absent from cities, most do still retain some endemic native species. Encounters with nature can therefore, vary across a spectrum of modification (see Figure 1.8). Children in different cities or parts of a city may be unable to access the full range of natural experiences present along this continuum. In highly urbanised, built or denuded environments, the most prevalent examples of the natural world may be the presence of cultivated plants, isolated tree species, animals with varying levels of domestication, such as dogs or cattle, or globalising species, such as the house sparrow. In city environments, it is often the presence of these global adapter and robust species that provide children with potential nature connections. In part, this is because these species are highly visible and tolerant of human presence, as demonstrated by the koi carp in Figure 8.2a. These koi carp are semi-wild, living in a park in Seoul, but their adaptability is shown in Figure 8.2b, where koi carp are a great attraction for children at Singapore airport.

Added on to the city form and its natural character is culture, which can both define and determine natural relationships. The ways that children's nature connections are portrayed through the back-to-nature movement reflects a particular way of seeing the relationship. Children are seen as removed from or above nature through what Malone (2016) calls object-subject binaries, an approach that she argues is not reflective of the myriad cultural understandings that inform children's very different relations to nature. She illustrates this through her study of children's relationships with dogs in La Paz, Bolivia. She demonstrates how children's relationships with these dogs challenge western-centric views of child-nature-dog relationships (Malone, 2016). In La Paz, dogs and children act as largely free agents in the landscape, coming together by mutual 'agreement'. Neither significantly determines the other's life course. Children may feed the dogs but can't usually guarantee food, so the dog must self-feed. Neither can the child control where the dog goes or even if the dog 'wants' to have a relationship with the child.

a

b

Figure 8.2 The koi carp shown in this picture from a park in Seoul are now widely distributed global adapters, familiar to city children in (a) ponds and (b) even in airports. However, in countries such as New Zealand, their adaptability is problematic, and they have become wild living in rivers and are now classed as an invasive pest.

In western societies, such as the UK, in contrast, the dog is wholly dependent on the family who 'owns' it, giving the dog-child a quite different power and interaction relationship. As with La Paz's dog-child interactions, children's relationships with nature and species wild, domestic and in between are complex and closely bound up with geography and culture. They represent complex ways of viewing not only the natural world but also children's positionality with regard to that world.

The importance of the cultural overlay similarly emerges from a study of rural and urban European Americans and Menominee (American First Nations people). They can live geographically in close proximity yet see their relationships with the environment quite differently (Bang et al., 2007). This study found Menominee children from as young as four years showed no reliable human-animal asymmetries (i.e., did not see a separate and hierarchical relationship with animals), compared to rural European American children. Even very young Menominee children demonstrated ecological reasoning compared to their urban counterparts and in the adult population:

> Almost all parents (and elders) expressed moral values and beliefs about the need to respect nature [but] European American adults mentioned the need to protect nature, implying a caretaker relation with the natural world. Menominee adults, in contrast, were more likely to stress that people are 'a part of' nature.
>
> (Bang et al., 2007, p. 13871)

What these examples show is the way culture acts as an important conduit for ways of seeing nature that differ not because of any significant difference in the nature present but because of the cultural overlay that urban-rural/First Nations/European American people place on nature. We continue the theme of culture through an exploration of different ways of viewing nature, including religious, indigenous views and some contradictory views towards nature.

Nationhood, religious and cultural belief systems

Nationhood: Different countries have different natural concepts integrated into their built form and their social practices as they relate to children and nature. We note that those we discuss are generalisations and that there will be considerable variation at the group and individual level. In understanding what is happening for children in cities, recognition of cultural practices present in the country matters. Culture is important in terms of national identity. Cultural practices and understandings are imperative if children's behaviours are to be understood and natural relationships encouraged and supported. A common example here that illustrates cultural difference is again the dog. Dogs are revered as a pet in many western cultures. Walking the family dog is a common way for children in Western Europe to experience the outdoors. In contrast, in some Asian cultures, dogs are considered more functionally as food, while some Muslims consider dogs as unclean and will not touch them. In India, in common with La Paz, dogs are integral elements of urban life, acting as companions but also as scavengers and potentially also a source of life-threatening disease (rabies being the most virulent). Interspecies relationships are complex and highly variable across place and culture. In India, cows and monkeys[5] are considered sacred and have free access to the city; however, their survival in cities is not always assured. Conflicts with people and competition for food can make their urban relationships difficult as well as revered. In Delhi alone, upwards of 30,000 cows roam free. For children living in Indian cities, the presence of cows, monkeys and snakes can be a prominent part of their nature connection, but understanding this connection requires understanding the status of these animals with regard to culture and religion.

Norway presents an example of a country where connection with nature is embedded in the national psyche. A study by Gundersen et al. (2016, p. 121) found,

> Norway is facing many of the same challenges of children's limited and often shrinking outdoor play territory as has been observed in many other western countries, with children mostly accessing places closer to home.

However, parental desire for outdoor play in nature remains strong; "most parents would like their child to spend more time in nature, and only 15% are satisfied with their child's current frequency of use of nature" (p. 119). His finding of dissatisfaction is interesting as some

> 97% of parents state that their children have access to forests within walking or cycling distance from home. When it comes to suitability for play, 88% state that their child, in general, has good or very good opportunities for play in nearby nature.
>
> (2016, p. 116)

Norwegian parents continue to believe in the idea of 'the robust' child and that spending time outdoors regardless of weather is a good thing (Nilsen, 2008). There is strong support for nature day-care centres, where spending at least two hours a day outside is the norm and an important part of children's agency and development. Given the harsh Norwegian climate, outdoor familiarity and knowledge is seen as important, as children need to be able to manage themselves to cope with climatic adversity. This idea is central to and reflects a historically rooted process of cultivating a national identity based on nature. For today's Norwegian children, where 80% of the population is urban,[6] wilderness and forests, the traditionally imagined outdoor habitats, are being replaced in their daily lives with woodlands and city parks. Nonetheless, the Norwegians' persistence in supporting an outdoor ethos should ensure that these urban replacement habitats as well as wilder habitats remain important conduits to children's nature connection.

Religion can be important in the same way, as belief systems are embedded in the national psyche. For Asian children, Daoist, Confucian, Buddhist or other beliefs can be significant. In Daoism, Dao and nature are indivisible entities with nature being a central part of the fabric of life (Miller, 2003). Confucians have divergent views on nature but have a rich vocabulary designed to describe the natural world, including terms such as 'qi' or vital force[7] (Bethrong, 2003). Nature is used for moral instruction where to be human is to be part of nature. Similarly, feng shui, with its central tenets of wind and water embodies natural elements and their relationship in daily life, where life occurs surrounded by natural elements, seen and unseen (Parkes, 2003). For Buddhists, nature was a central part of Buddha's own life:

> Everything arising out of Dharma, everything born from Dharma, is what we mean by 'nature.' This is what is absolute and has the highest power in itself. Nature has at least four aspects: nature itself; the law of nature; the duty that human beings must carry out toward nature; and the result that comes with performing this duty according to the law of nature.
>
> (Sponsel and Natadecha-Sponsel, 2003, p. 355)

This centrality of nature in belief systems is not peculiar to Asian belief systems but rather is typical of a range of beliefs, including Islam and Christianity. For Christians, the Bible begins with the story of creation, in order – light, water, sky, land, vegetation, living creatures and then humans, except that humans were given authority over nature. Beliefs about the need to connect with nature can be seen in cultural representations of nature. Zen Buddhism reflects these values in its gardens and symbolic plantings and patterns. Chinese gardens, such as the ones in Figure 8.3, are small-scale intimate gardens. A scholar's garden may follow a

Figure 8.3 The Chinese Gardens in Dunedin, New Zealand, are a popular family attraction and reflect a culturally determined relationship between built and natural elements.

Figure 8.4 In Singapore, children are encouraged to feed the turtles at the Live Turtle and Tortoise Museum at the Chinese Gardens in Jurong Park.

traditional design, where people could experience in miniature an ideal landscape that allows escape into nature. In Singapore, the Chinese Garden includes a tortoise and turtle centre where children are actively encouraged to interact with the turtles by feeding them (Figure 8.4).

Belief systems can play an important role in children's relationship with nature and their attitude towards it. However, maintaining culturally appropriate attitudes and beliefs can be challenging when the environment in which the child is placed is antithetical to nature. This would be the case for a child growing up in a high-rise high-density Hong Kong living environment, such as in Figure 8.5. Changes in emphasis in children's lives can also make it difficult for children to connect with nature. In Hong Kong, life pressures, such as the pressure to succeed educationally in an uncertain future, are becoming the central tenet and a driving force in child-rearing for many parents. Educational and 'enrichment' activities leave little time for children and their families to engage with the outdoors (Karsten, 2015). Several researchers have noted a process whereby education and 'future-oriented' activities displace time for relaxation and play for many children, leaving little if any time for nature. Daily reality then can be at odds with nature as it is enshrined in higher belief systems. Dissonance between natural belief systems and the built

Figure 8.5 Residential area in Hong Kong where children would struggle to connect with the natural environment.

environment can, however, be mitigated. While the focus on education is also strong in Singapore and high-rise is the norm, living environments have been designed to embrace natural features in accordance with the national greening plans for the city-state. The Singapore and Hong Kong examples occupy different points on the nature continuum (Figure 1.8). Singapore is in the middle of the continuum with urban Hong Kong near the start. Singapore shows what may be possible in terms of greening (albeit an artificial greening) and how there can be a place for engaging with nature as a part of an urban environment and lifestyle.

*Celebrating culture and natur*e: Culture allows for celebrations of nature, though not all cultural celebrations necessarily invoke positive messages about the nature connection. Questions have been raised about the acquisition and management of elephants in Asian countries that take part in religious and other celebrations. There are some festivals which involve treatment of animals that is contrary to any positive human-species relationship. This is illustrated in the

nzherald.co.nz

Search for content...

National Opinion Business Tech World Sport Entertainment Lifestyle

National NZ Herald Focus Crime Politics Opinion Employment Health Insights Educati

School criticised over possum throwing

10:44 AM Thursday Sep 23, 2010

Animal Welfare | Manawatu | Manawatu-Whanganui | ...

SHARE:

A rural Manawatu school which held a possum throwing contest has drawn SPCA criticism and upset many in the community.

Children at Colyton School last week held possum carcasses by the tail in a contest to see who could throw them the furthest, the *Manawatu Standard* reported.

File photo / Bay of Plenty Times

Palmerston North SPCA centre manager Danny Auger said the school was not breaking any laws, but that did not mean children should be encouraged to mistreat dead possums.

"It's morally wrong to throw a dead animal around. It's about time that people wake up and smell 2010 and realise that these sorts of things shouldn't be happening."

Mr Augers was worried that children were not being taught to respect dead animals.

He would visit Colydon School to ask them to stop holding such events.

The school refused to comment.

- NZPA

Figure 8.6 Not all cultural events are respectful of nature as indicated by Colyton School's possum-throwing competition.

Source: www.nzherald.co.nz

newspaper report extract on a New Zealand school's possum-throwing competition (Figure 8.6). However, cultural celebrations can help children respect nature and animals. Each year in October, the city of Vina del Mar, Chile, hosts a bird festival attended by families as well as ornithologists. In Dunedin, the church bells ring each year when the first royal albatross returns to the mainland to the breeding colony. Less tangible are representations of the spirit of an animal in celebrations. The lion dance is celebrated in Indonesia, Vietnam, Japan, Nepal, Tibet and China, despite the lion not being a species found there currently or historically. By taking part in such ceremonies, children do nonetheless imbibe

Figure 8.7a Children celebrating Chinese New Year through the Lion Dance, Dunedin, New Zealand.

Figure 8.7b Dunedin is home to the world's only mainland breeding site for royal albatross. The special significance of this bird is acknowledged in a number of ways, including ringing of church bells when the first one returns for the breeding season and in its inclusion in the annual Santa parade.

the spirit of the animal and have enhanced recognition of the species, albeit in a very stylised form (Figure 8.7a). In Dunedin, New Zealand, iconic species, such as penguins and albatross, form part of the annual Christmas parade (Figure 8.7b).

Plants also have symbolic relationships linked to culture, form part of cultural festivals and have particular meanings. In Seoul, families can be seen out enjoying

Figure 8.8 Schoolgirls in Seoul admiring cherry blossoms in the Gyeongbokgung Palace Gardens, Seoul.

the cherry blossom season with visits to places such as Namsang Park, which is known for its spectacular cherry tree blossom (Figure 8.8). Festivals or traditions around plants often serve to emphasise seasonality. While the examples so far include reference to cultivated and natural landscapes and symbolic nature relationships, culture also influences relationships to specific species. Certain species are seen as having particular roles and cultural meaning. In New Zealand, for example, birds such as the fantail (piwakawaka) and the owl/morepork (ruru) feature strongly in Māori oral histories. These animal stories have particular roles and are referred to in sayings. Some examples demonstrate this:

1 The *ruru* (owl/morepork) acts as a *messenger for the spirit world*;
2 Departing spirits are compared to the migratory godwit or kūaka in the saying '*me he kāhui kūaka*' (like a flock of godwits);
3 *Pūkeko* (purple swamp hen) are known for their bold *scheming and determination* – in the past they raided gardens for kūmara (sweet potato) and taro. A stubborn, annoying person was compared to the bird, and was said to have pūkeko ears (taringa pākura);
4 In one traditional story, the *piwakawaka* (fantail) *caused Māui's*[8] *death*, so it is seen as unlucky and a harbinger of death when seen inside a house;

5 *I whea koe i te tangihanga o te riroriro* (grey warbler), *ka mahi kai māu?* Where were you when the riroriro was singing, that you didn't work to get yourself food?;
6 *He rite ki te kōpara e kō nei i te ata.* Just like a korimako (bell bird) singing at dawn. Great singers and orators were praised by being compared to the korimako, a beautiful singer;
7 People who are eloquent speakers, talkative or quarrelsome are often compared to the tui.[9]

(Keane, 2015)

Māori tradition is replete with stories and sayings that have meanings and connections, not just to the bird and its character but also to the bird's past and current role in Māori culture and history. Some of the birds referenced in 3 to 7 would be recognisable to most city children today, being present in gardens and local parks and bush. These birds enable children to develop a connection not only to nature but to wider culture and history through sayings and stories. Stories and beliefs that connect nature and culture are also evident across cultures. In English, we use the expression 'dead as a dodo', a saying that most English-speaking children are familiar with and which evokes the susceptibility of species to extinction. The European robin graces numerous British Christmas cards. Why the robin? Possibly because it has a bright-red breast and so is very noticeable, but also it is not as wary as most birds around people, and in winter (Christmastime), when food is scarce, it is frequently seen around homes. The robin also appears in Christian folklore with roles relating to Christ's birth. More prosaically, the first postmen who delivered early Christmas cards in the 1800s wore red coats. Feeding birds in winter, including robins, is a strong tradition in Britain (over half the adults in Britain feed birds according to the Royal Society for the Protection of Birds), and the robin is possibly the most widely recognisable bird for British children. The robin was voted Britain's most popular bird in 2015, taking 34% of the vote for the 10 birds in the voting (*The Guardian*, 2015). In New Zealand, the robin also appears on Christmas cards, but it is the European Robin rather than the New Zealand robin, which like many other endemic birds has not made the transition to urban living. It is only recently that New Zealand's own native fauna and flora has started displacing European flora and fauna on cards, reflecting a move away from a strong colonial influence.

City nature is both an adapted nature and one that is culturally constructed. Some nature features at one end of the spectrum are physical constructions and artefacts, as in the Japanese Zen gardens, Chinese gardens, the formal parks following British traditions or koi fishponds originating in Chinese cultures. At the other end of the spectrum are less adapted, more natural moving towards more unmanaged wilderness landscapes. Within all urban landscapes, there are embedded cultural manifestations and different ways of seeing and valuing the natural environment that is present. The example of the Chinese garden in Figure 8.3 demonstrates this. Rocks are carefully selected and positioned. Plants are carefully placed and cultivated. The water feature too is highly regulated. These built and natural features are embedded with cultural and spiritual significance that would be understood by Chinese children and their families.

Identity and belonging

For city children growing up in cultures with a long established urban history, such as the UK, Europe, China and parts of Asia and the United States, there has been a longer process of acculturation. In places with a long urban history, the urban form has grown in a more gradual, organic way with people and the species found there adjusting to each other over an extended time period. The urbanisation process reflects cultural ways of living and the landscapes that people in those cultures value. For example, in the US and UK, the suburban garden has developed its own place in society with the garden itself following a recognised cultural style of gardening. However, for children who move into a new society, or where an urban landscape and way of living is imposed on their living space, there may be no opportunity to gradually adjust, and the changes experienced are radical and rapid.

Nature is not universal in its forms or interactions, as already discussed. As children move to different spaces, rural to urban, lower to higher socio-economic, low-density to high-density living environments, their nature relationship changes, and they need to adjust as a consequence. For example, a child moving from a terraced street in the UK with its close proximity to the public space of the street would gravitate to the local parks rather than private gardens for play. In suburban Australasia, domestic gardens, which tend to be larger, are more usually the play space rather than the street, which is predominantly a traffic space.[10] A child moving from a high-rise apartment in a Chinese city would find the adjustment in moving to suburban Australasia even greater. Nationality, culture, religion and even politics play a part in shaping this relationship. Under the apartheid system in South Africa, the more affluent White population had gardens characterised by greater species richness and abundance, with more non-native species.[11] In contrast, the lower-income Black African population had smaller gardens with fewer species but more native species, including species that could be used for food and medicine (Nilon, 2014). What if children find themselves with cultural understandings that are out of sync with the dominant nature-relationship ethos or if children do not comprehend that dominant ethos? A child from the affluent white suburb in apartheid South Africa would be unlikely to recognise or value the plant species in the lower-income black child's garden and be unlikely to recognise their medicinal properties.

A relationship with nature in the natural places available and with the species present necessitates that the child feels comfortable in these spaces and with these species and can read their associations. For example, an urban woodland often lacks any direct indication of who it belongs to; is it public or private, do you pay or not pay, what activities do you do when you get there, how long will it take to get to the other side and where do you exit, are there any facilities in the woodland, do you need special clothing (such as shoes for muddy places), and are there dangerous species (such as particular snakes or spiders)? If a natural place gives no such information, then when encountering an urban woodland for the first time, children will feel hesitant to continue with the encounter. If there is no experience the child can draw on from friends or family, then the urban woodland may be avoided. This type of dilemma can be present for many children newly arrived in a different environment from the one to which they are acculturated. For the children amongst the estimated 1.1 million, mostly Syrian refugees entering Germany in 2015, understandings of an approach to their new nature and natural spaces will need to be learnt. For these children, many traumatised by war and conflict, their priorities could also be different. Further, given the transitory nature of refugee life, the sense of understanding and belonging necessary to use and explore natural places may be difficult to acquire. These children's understandings of nature may have been acquired with reference to a very different natural and socio-cultural framework and one that may not translate easily to their new environmental context.

Adams, though not a refugee, describes her sense of place in a transnational community where children live in a hybrid space. Her childhood experience combined her Caribbean heritage and her physical living space of New York. She recounts how different Caribbean communities gather in parks to reconnect socially and culturally. Another way to connect is through community gardens, which mimic the kitchen gardens characteristic of Caribbean communities as places to grow Caribbean foods (Adams, 2013). The New South Wales Service for the Treatment and Rehabilitation of Torture and Trauma Survivors, Australia, has initiated a number of community garden projects for refugees in recognition of the benefits such gardens provide for families and survivors. Similar benefits were identified in Latino community gardens in New York City:

> Children under 13 years old, although not well represented among gardeners and garden members, are the age group that visits the gardens most often (40% of total garden constituency, relative to 30% of total garden constituency for senior citizens and 22% for working adults). Most gardeners are happy to teach young people about gardening and many gardens have toys and activities geared toward children . . . Another garden, Villa Santuce, hosts tours for groups of children with disabilities.
>
> (Saldivar-tanaka and Kransky, 2004. pp. 7 and 10)

For these Latino children, community gardens provide a place where they can connect not only socially to their cultural and ethnic group but also in an outdoor space that is welcoming and understood – a contrast perhaps to the woodland example. The notion of spaces being welcoming or safe has been identified as an important factor in minority ethnic children's use of outdoor spaces. Karsten's study (1998) of segregation in children of different ethnicities in Amsterdam found clear evidence of differentiation by ethnicity in the use of outdoor spaces, concluding that the children who played outside least were children of Surinamese/Antillean descent living in the Bijlmer neighbourhood. A more recent study also focusing on immigrants in the Netherlands found "landscape preferences differ significantly between immigrants and native Dutch people" (Buijs et al., 2009, p. 113). While native Dutch are strong supporters of wilderness (albeit following a Dutch perception of wildness[12]), this is not the case for immigrants. Immigrants have quite low preferences for wild and unmanaged landscapes, like marshes and dunes, and a preference for more functional, managed landscapes. However, second-generation immigrants, born in the Netherlands but raised in an immigrant culture by parents originating from Turkey or Morocco, were found to be in a phase of acculturation in their views of more natural Dutch landscapes.

Being an immigrant also impacts on access to green spaces. A study of children in the US examining their activity rates found "Lower-income groups and racial and ethnic minorities have limited access to well-maintained or safe parks" (Active Living Research, 2011, p. 2) giving them reduced access to natural spaces. This finding led the Active Living Research organisation's[13] report to recommend prioritisation of "lower-income communities and communities of color when maintaining and creating new neighborhood parks and pocket parks, open spaces and other destinations for recreational physical activity" (2011, p. 12).

The number of indigenous children in cities is growing. The Aboriginal population[14] in Canadian cities is growing substantially faster than the non-Aboriginal population (Guimond et al., 2009). For indigenous children growing up in urban environments, there can be particular challenges. The urban environment can be one that is at odds with the children's cultural inheritance and ways of seeing the world and nature. Indigenous children are generally impacted by low socio-economic circumstances. In New Zealand, Māori are over-represented in every negative social statistic, including health, education and income. As a consequence, "the depth and persistence of poverty in some communities means many are likely to get caught in the cycle of low educational attainment, unemployment, low-income, inadequate housing, and persistent ill-health that has become the norm in many families" (Māori Affairs, 2012, p. 9). Māori families can find it difficult to translate the social and cultural safeguards present in more traditional ways of living into urban life. Traditionally children were kept busy fishing, snaring birds, forest food harvesting and helping in the garden (Jenkins et al., 2011). Children had a special place in society, being seen as closely linked to the spirit world, which accorded them safety and respect, and were seen to need turagawaewae 'a place to stand'. For children, the disconnect from cultural belief systems and their place in the natural world, which can be most evident in urban Māori, has impacted negatively on their well-being.[15] There are more positive stories, and iwi are closely involved in the Orokonui Eco Sanctuary outside the city of Dunedin. Children are involved in many of the sanctuary's activities. Children take part in the formal iwi exchange and ceremony (Pōwhiri) that occurs when species are translocated and released into the sanctuary. This photo (Figure 8.9) is taken at the release site. One of the boys wore a traditional cloak/korowai and was at the forefront during the handover of the South Island robins-toutouwai and their release into the sanctuary.

For Australian Aboriginal peoples, such as the Arrenrte people, "'Dreaming stories' . . . handed down from generation to generation through storytelling, song

a b

Figure 8.9 Children from the local iwi play an important role in the ceremonies and exchanges that go with the release of threatened species into their iwi's local eco sanctuary. The children can live in the nearby villages and city but come to the sanctuary to take part in the release.

and artworks . . . explain how the different parts of the country and its inhabitants (humans, mountains, rocks, rivers, plants and animals) came into being" (Taylor, 2013, p. 94). Affrica Taylor, from whose book this quote is taken, talks of how her experiences with the Arrernte "quickly unravelled any western conceits I may have held about the universality of the individual child . . ."(Taylor, 2013, p. 95). Taylor then poignantly goes on to reflect on the fact that most Arrernte children now live on the urban fringes, experiencing locational and cultural marginality from both traditional and modern western worlds. Such marginalisation through the vicissitudes of postcolonial repression and alienation, challenges the rich nature relationship these children should rightly inherit. However, to see children as mere alienated victims is equally incorrect as through "the adaptive capacity of the Arrernte culture . . . its survival, these same children-animal relations produce newly constellated kinship relations and contribute to the creation of continually reimagined past-present modern 'hunter gatherer' subjectivities, lifestyles and practices" (Taylor, 2013, p. 102).

Similar challenges are noted in a different study of aboriginal children in Cherbourg, a small town in Australia. Here, Kreutz (2014) found that what she calls 'cumulative stressors', such as crowding, alcohol-fuelled violence and socio-economic disadvantage, directly influence the degree of child-environment congruence. While the children had few restrictions on their independence, the environment itself did not feel safe for the children socially and sometimes physically (blue-green algae precluded swimming in the creek). As a result, "the natural environment surrounding the community town did not present the children with the same opportunity-rich environment that was experienced by their forebears" (Kreutz, 2014, p. 73). Re-engagement with cultural practices in the local environment is seen as essential to recovering children's engagement with their natural environment. Indigenous children can come from a cultural background rich in beliefs and practices connecting them to the natural world. The disruption to this connection when they move into urban environments can be felt particularly deeply and impact in significant and negative ways on their well-being. The future reality for many indigenous children will be an urban one; the challenge is how to sustain children's beliefs, practices and sense of who they are in the built and natural worlds and in worlds that are not currently supportive of either their physical or emotional needs.

Accessing nature for children with disabilities

Another group whose needs are not well met are children with disabilities. UNICEF estimates some 93 million children, or 1 in 20 of those aged 14 or younger, live with a moderate or severe disability of some kind (UNICEF, 2013). Nature experiences may have specific benefits for children with disabilities. There is considerable evidence that nature experiences can assist in reducing some atypical behaviours (e.g., those that characterise ADHD), reduce stress levels and enhance cognition through improved focus (von Benzon, 2011). Such experiences can be especially important for children with intellectual disabilities. Overall, there is limited experience of providing for the nature needs of children with disabilities, especially for children with physical disabilities or multiple disabilities. Where disability has been addressed in relation to the environment, it invariably relates to matters of physical accessibility, especially in relation to more formally designed environments, such as classrooms or playgrounds, rather than access to informal or wilder natural environments.

For many children, exclusion rather than inclusion is the norm, especially when it comes to accessing the outdoors. Nature and natural places can provide opportunities where all children can come together and experience nature. The outdoors provides opportunities for inclusion and interaction, where children with disabilities connect with other children (Woolley et al., 2006). In playgrounds, swings are now sometimes provided for children with disabilities, and in some cases, they are placed behind a fence, exaggerating separation. Good intentions, therefore, can inadvertently exaggerate difference and separation if done badly. The outdoors cannot easily be fenced off, and the range of interactive elements – birdsong, trees of different levels of climbability, running water of different depths, sand and earth – all provide interactive and experiential affordances for all children. It is not possible to provide for all disabilities in all natural places, but the needs of children with disabilities can be met by ensuring that there are different types of spaces available to meet their needs. These spaces need to be accessible from their homes. Resources may also need to be made available to provide safe places; these can be home-based, neighbourhood-based or further away places for less frequent visits. Provision needs to be more than providing for disabled access and should take into account sight, hearing and other impairments and provide for the needs of children with intellectual impairments, such as autism. Provision should be made for children whose needs are complex and may need medical support equipment to be able to be taken on site (Woodgate et al., 2012). Taking into account and enabling inclusion of disabled children's needs in the design of outdoor spaces needs to be a normal part of the project development and participation process.

Children with a disability or special needs are only individuals in so far as they have particular behaviours or needs; in all other ways, they are part of families, communities and societies. If these children are unable to access nature experiences and opportunities, this can directly impact on and restrict the ability of the family to experience nature. Inclusive accessibility to nature experiences and provision of the supports necessary to enable children with special needs to access nature has direct benefits not only for the children but also for their wider family and friends. For example, a beach that has limited access or no access for a child with a disability can preclude all the family accessing that beach. A Canadian study interviewed parents of children with complex care needs.

The study findings reinforced the notion that for families of children with complex care needs, "Places, spaces, activities, and situations had to be easy to move in and out of in order to be considered accessible . . . and [they] wanted to be able to access the same places, spaces, activities, or situations that 'able bodied people' have access to" (Woodgate et al., 2012, p. 1915). One father described the limitations as follows:

> We can't take him to any outdoor activities because we can't rely on pathways and sidewalks to be passable via wheelchair. We can't take him to the nature retreat, for example. We can't move him around easily, even with the wheelchair van we can't easily get him from where we would have to park to the place, and it's very difficult, physically demanding to push him across the, the landscape. It's impassable (Father 14).
>
> (Woodgate et al., 2012: p. 1917)

A consequence of inability to access the outdoors can mean home becomes the only accessible landscape. Further, even where families can access the outdoors, compromises may have to be made, and Woodgate et al. give the example of one family who "opted for a three-day camping trip after acknowledging their preferred seven-day camping trip was not possible" (2012, p. 1918).

While families experience difficulties as they try to provide meaningful encounters for their children, significant barriers impact on children in the school environment. A European Commission study on disability found, from their survey of

60 organisations, "Over 73% declare that no support is provided for extra curricula activities organised by the school, undermining the possibility of integration of children with disabilities within their school environment" (European Disabilities Forum, 2003, p. 31). This means children will miss out on most outdoor activities and trips.

Safety is a dominant issue for families and children with disabilities. In a natural environment, it is inevitable that some degree of risk will be present. When parents in Woodgate et al.'s study were asked about risk, they noted, "While safety was always considered, some parents expressed that sometimes it was important to allow children to partake in activities that involved some risk" (Woodgate et al., 2012, p. 1915). It can be difficult to get the balance right. Von Benzon relates how even where children are taken (in this case autistic children) to natural places, they were not allowed to enter the woodland or use the fields, slopes and water but had to take part in organised activities in clearings separated from the woodland by a fence (2011, p. 1033). She concluded that "the environmental opportunities offered to participants were characterised as being highly managed, affording the young people few opportunities for independent exploration of natural environments" (2011, p. 1035). Health and safety concerns can be especially problematic and restrictive for children with disabilities. A child with a physical disability, such as impaired vision, could have their ability to go into natural spaces, such as forests, on a school outing curtailed if the school was compelled by health and safety legislation to remove all chance of accidents or risks for children with disabilities. In such a case, the school may feel the visually impaired child has to be withdrawn from the outing as they cannot guarantee a risk-free environment, thus denying the child the opportunities the natural encounter afford to her or him and indeed to the other children who cannot be guaranteed a risk-free environment either. Enabling children with disabilities requires rethinking how to manage risk so that the outdoors can be safely accessed and risks reduced but not to the extent that the child becomes disconnected from the natural world.

The example of the residential schools in Box 8.1 provide clear evidence that an approach where the outdoors and access to nature is seen as nurturing and to be engaged with rather than as unsafe and to be avoided is not only possible but preferable. Most of the examples used here and the example of the residential schools relate to children being taken to special, usually rural nature settings. While these experiences are valuable, they need to be supported by similar and accessible nature experiences in the home environment. The challenge is to provide children with the skills to enable them to safely negotiate more natural environments not to eliminate exposure to potential risk.

There is a shortage of nature opportunities appropriately resourced for use by children and families of children with disabilities in urban areas. Schools can be a good way of providing such opportunities. Regency School in the UK caters for students aged 11 to 19 years with a range of additional needs, including moderate, severe and profound learning disabilities; physical and medical needs; and sensory impairment. Working with nature in a range of ways is an important part of the

Box 8.1 Special education with a nature focus

The Steiner Waldorf organisation provides a number of residential schools for children with special needs. These schools located in rural environments stress the importance of providing special education with a nature focus. The following extract is from the school web site for Ringwood School, UK. The school specialises in providing for children with autism and communication difficulties, severe and complex learning disabilities and challenging behaviour. They emphasise the importance of nature in their publicity about the school: "Set in a stunning sustainable and eco-friendly 50-acre site close to the New Forest and Dorset coastline, the natural surroundings of our school and college provide a rich, enabling and therapeutic learning environment for students, aged 6–25, who have a special educational need." Further, they emphasise, "The Outdoor Curriculum offers a therapeutic and educational programme to children and young people who are hard to reach and who may have experienced difficulties with a more formal approach to education and rigid timetables. This approach has been very successful with children and young people who have been out of school for long periods of time or who found it difficult to adapt to a classroom setting due to behaviour difficulties". Their aim is to "help each pupil integrate their sensory experiences through encouraging exploration of the natural environment." Thus unlike mainstream education, which can find the outdoors too difficult an environment to work in with children with special needs and thus preclude the children from participating in outdoor education and experience, these schools see the outdoors as a fundamental component of the children's educational and social environment.

Taken from information provided about the Shieling School Ringwood UK based on Steiner-Waldorf principles, www.sheilingschool.co.uk/home.

school curriculum as described by one teacher commenting on a project to create a garden for the Malvern Spring festival:

> Gardening is so levelling for students with special needs, it's very inclusive, says Sarah Rudge, a teacher at the school. It takes a long time to see things develop and grow, but having patience is a really good life skill, she adds. While some children will concentrate on digging earth and planting seeds, others will be calculating the dimensions of the garden or researching which herbs the Romans grew and what vegetables they ate.
>
> (*The Guardian*, 2016)

On the same theme, the same newspaper article quotes Dr John Crosbie, former director of the UK charity The Calvert Trust, which provides outdoor activities for people with disabilities. On the benefits of the outdoors, he states, "Outdoor activity requires students to adapt to unpredictable weather and the changing seasons,

Figure 8.10 Catching the waves during the king tide is an opportunity for children to experience the power of nature. The Dunedin Esplanade is accessible to all children, allowing them to experience this seasonal event together. Unfortunately, the same cannot be said for the beach below, which remains inaccessible to children with disabilities since the access ramp was destroyed by waves soon after construction.

and this can be another positive challenge" and "The ability to try and relate to situations is a real challenge for children on the autism spectrum, and so is being flexible with their view of order, timescales and routine; being outdoors challenges this." The Calvert Trust again focuses on "going into the outdoors through dedicated centres" and thus provides one-off but highly valuable activities. This emphasis on the special outings and experiences again underscores the need to also provide for localised, easily accessible urban-based opportunities that make the benefits of outdoor activity available as part of normal, everyday life. These everyday experiences are vital to embed the special and to build on them through other more reachable life experiences (Figure 8.10).

Nature as threat

The notion of nature as good, welcoming and benign is at odds with many children's experiences. For some children, lack of contact or understanding explains their apprehension about engaging with natural places, but for others, their negative views are based on concerns that are real and valid. Natural places can be places children can fear to go because of who does or could use the places – older young people, drug users or others whose presence makes children feel unsafe. The messages given by the physical landmarks around a natural site can give out strong messages. The photos (Figure 8.11a – 8.11e) taken some years ago at Bog Meadows Nature Reserve, Belfast, reveal this. Described by Ulster Wildlife (n.d.) as "An oasis for wildlife and people, in the heart of Belfast city", a child going to the site then would have had to pass through intimidating walkways with glass, graffiti and defensive fencing, all of which give very negative messages compared to that given by the 'oasis' beyond. The site also carried the history of being located in an area that was severely impacted on during the "Troubles" or Northern Ireland conflict.[16] The site is now an important wildife site and an increasing focus for community activities.

For children in South Africa, the real and potential threat of violence outside the home is a major constraint on their outdoor lives. The legacies of apartheid and its associate civil tensions have created a legacy of violence and fear for many urban residents (Richter, 2001). This fear was evident in a study of children who lived near Western Joubert Park in Johannesburg's inner city who were fearful of using the park (Swart Kruger and Chawla, 2002). Parents were particularly concerned about the safety of their girls so restricted their outside play. Elsewhere in Johannesburg, Malvern Park is avoided, as it is dirty, has litter, and adults go there to drink. In the suburb of Riverlea, some boys do play on the waste dumps (mining waste) though their parents forbid it. Also in Riverlea, some children play on a small hill 'the

a

Figure 8.11 The Bog Meadows Nature Reserve, an urban wildlife oasis, in the past necessitated passing by intimidating fencing, graffiti and glass-strewn pavements to get to the meadows and wetlands.

b

c

Figure 8.11 (Continued)

d

e

Figure 8.11 (Continued)

Buljtie', as the park is unsafe, but they do have to deal with young people drinking, smoking dope and having sex. In general, open field and natural spaces are avoided as dangerous, as people have suffered violent attacks there. In another suburb in the Johannesburg study, Pimville children are fearful of the parks and open veld (grassland) so in preference gravitate to a lawn near a local garage. While South African cities have particular problems regarding violence, interviews with children living in Suva, Fiji, showed an awareness of related problems of safety around crime with children avoiding certain places, but the avoidance was strongly related to night-times. Children related that they generally felt safe during the day and often used school fields as play spaces, as these were seen as safe (Freeman et al., 2015).

The outdoors can be a place of danger for children living in conflict situations. A study conducted in 1994 in Palestine found that mothers rated the most serious danger for their child as being beaten by soldiers, followed by being shot, teargassed and detained. The order varied depending on the age of the children (Kostelny and Garbarino, 1994). Some 23 years later, in another study, Giacaman et al. (2007) report that of the young people in the study living in cities, 78% were exposed to shelling in the neighbourhood, 75% to teargas, 36% body searches, 50% house searches, 66% the sound of bombs, 45% the house being shot at, 50% a friend or neighbour being arrested, 35% a stranger being killed and 7% a family member being killed. These are selected examples and resonate for young people in towns, villages and camps in Palestine. In conflict scenarios such as in Palestine, freedom is severely curtailed, and where conflict occurs over prolonged periods, as in Palestine, nature relationships decline. For example, many Palestinians are denied access to their traditional lands on which they grew olives and other crops (United Nations Office for the Coordination of Humanitarian Affairs, 2012).

The outdoor natural environment itself can also be problematic. Water can be polluted and land contaminated or just unpleasant, as was the case for the children we interviewed in Suva. They explained how they disliked mud, as when it rained, they had to walk to school through a swampy area ending up with muddy shoes and clothes. In India and countries with heavy rainstorms, rain can bring with it danger from flooding, stagnant water, waterborne disease and animals, such as snakes or crocodiles who come in with the newly formed lakes and rivers. Natural disasters and associations with danger can make children wary of natural places. Low-lying cities and settlements are in danger from tsunamis. As many as 100,000 children were displaced by the 2011 Japanese tsunami, and many of those who lost their lives were children (Save the Children, 2011). Children in temporary housing had little, if any, play space (Kinoshita and Woolley, 2014) with some 260,000 people still living in temporary houses three years after the disaster. The housing developments had not provided for play space in their design. Concerns around radiation levels post-disaster from the Fukishima plant also meant that many parents and schools were restricting their children's outdoor play. A number of indoor play spaces have been constructed, 64 alone in the Fukushima Prefecture (Kinoshita and Woolley, 2014).

Similar concerns about being out and about were noted after the Christchurch earthquake in New Zealand. The two major earthquakes of September 2010

(4/9/2010) and then February 2011 (22/2/2011) were followed by an extended period of aftershocks lasting well over a year. The ever-present threat of further earthquakes created a high level of parental and child anxiety that meant many children were not allowed or were reluctant to venture far from home. Many parks and sports fields were severely damaged by the quakes so they became unsafe, very uneven, and many experienced liquefaction. After the quake, some children moved many times, including out of Christchurch (Freeman et al., 2015). This severely disrupted children's social networks, which have been identified as one of the key factors in supporting independent mobility, a prerequisite for outdoor play (Mikkelsen and Christensen, 2009). Disasters and lack of safety can, therefore, severely impact on children's ability to seek out and use natural spaces.

Nature contact a complex web

Children's connection with and use of natural space is determined by two key factors: first, the availability of green space and a perception that the space is inviting and safe to use, and second, the space is one that the children's culture, socialisation and position in society supports the use of. Children's relationship with nature and the natural spaces present needs to be in accord with social and cultural supports. This support is clearly present in Norway, but even there, adjustments are occurring, as the wilder places at the heart of the Norwegian relationship give way to the more managed and functional natural places available to city children. However, where children are traumatised, as in Palestine, open spaces are seen as unsafe, and for different reasons, open space is also seen as unsafe in Johannesburg. More culturally and socially aware nature connection strategies will be required where children lack the knowledge, cultural connection (as when they are refugees or in transit) or individual factors, such as their gender, preclude access. It is not enough for natural open spaces to be present for children to use and relate to. A far more complex web of relationships is implicit and present, and it is this web that this chapter addressed and attempted to make explicit.

Notes

1 Of European descent, *Pākehā* is the Māori word for NZ Europeans.
2 Sustainable Singapore Blueprint 2015, Singapore Green Plan 2012, National Biodiversity Strategy and Action Plan 2014.
3 The government recognises the relationship between an attractive green environment and economic investment.
4 Fiji's two main population groups are Indigenous Fijians and Fijians of Indian origin, IndoFijians.
5 Hanuman the monkey god is an important Hindu deity.
6 https://knoema.com. World Data Atlas Norway Demographics.
7 New Zealand Māori have a similar concept 'Mauri', the life force often used in relation to water.
8 A demigod and traditional hero important in many stories, e.g., he hauled up the North Island of New Zealand from the ocean, his canoe became the South Island and he snared the sun to slow it down and found fire.

9 An endemic and distinctive New Zealand bird described as follows: they are usually very vocal, with a complicated mix of tuneful notes interspersed with coughs, grunts and wheezes, www.nzbirdsonline.org.nz/species/tui.

10 This was the scenario faced by one author's children when moving from UK to New Zealand. The children lacked an understanding of how to approach friends for play in the larger fenced private suburban gardens, where homes were 'hidden' from the street, as their experience was of going to the street or local park to play.

11 The 2011 South African census uses the following population classifications, White, Black African, Coloured, Indian or Asian and other. Census 2011, Census in brief, Report No. 03-01-41, Statistics South Africa, http://www.statssa.gov.za/census/census_2011/census_products/Census_2011_Census_in_brief.pdf

12 The Dutch preference encompasses habitats and landscapes that to many would be more accurately classed as managed nature, i.e., polders, woodlands which contain natural elements but are themselves part of a larger managed landscape. This concept aligns more to arcadian views of nature rather than true wilderness.

13 A research organisation administered through the University of California that translates and disseminates evidence to advocates, policy-makers and practitioners aimed at preventing childhood obesity and promoting active communities.

14 The three main groups are Inuit, Métis and First Nations.

15 Māori and Pasifika children have 2 to 3 times the child abuse rates recorded for European New Zealand children, and the rate is growing, similar disparities in poverty rates are also evident (Wynd, 2013).

16 During the conflict, according to the book *Lost Lives* (2006 edition), 3,720 people were killed as a result of the conflict of which 7% were children. During the conflict, it would have been difficult for children to access spaces such as Bog Meadows. In 1998, Ulster Wildlife Trust received funding from a Peace and Reconciliation Fund to safeguard the land. Bog Meadows access has improved since these photos were taken and has been an important contributer to the peace and community building process.

References

Active Living Research. 2011. *Research Synthesis, Building Evidence to Prevent Childhood Obesity and Support Active Communities*. www.activelivingresearch.org

Adams, J.D. 2013. Theorizing a sense of place in a transnational community. *Children, Youth and Environments* 23(3): 43–65.

Aronson, M.F.J. et al. 2014. A global analysis of the impacts of urbanization on bird and plant diversity reveals key anthropogenic drivers. *Proceedings of the Royal Society of London* 281: 20133330.

Auckland Council. 2016. www.aucklandcouncil.govt.nz/EN/planspoliciesprojects/reports/Pages/censusinaucklandhom

Bang, M., Medin, D.L., and Atran, S. 2007. Cultural mosaics and mental models of nature. *PNAS* 104(3): 13868–13874.

Bethrong, J. 2003. Confucian views of nature. In H. Selin (Ed.), *Nature across Cultures: Views of Nature and the Environment in Non-Western Cultures*, Kluwer Academic Publishers, pp. 185–209.

Buijs, A.E., Elands, B.H.M., and Langers, F. 2009. No wilderness for immigrants: Cultural differences in images of nature and landscape preferences. *Landscape and Urban Planning* 91: 113–123.

Ellen, R. 2003. Variation and uniformity in the construction of biological knowledge across cultures. In H. Selin (Ed.), *Nature across Cultures: Science across Cultures: The History of Non-Western Science*, vol. 4. Springer, Dordrecht.

European Disabilities Forum. 2003. Disability and Social Exclusion in the European Union Time for Change, Tools for Change. Final Study Report, European Disability Forum European Commission.

Freeman, C., Gollop, M., and Nairn, K. 2016. Disasters, displacement and disruption: Children and young people's experience of spatial change following disasters. In N. Klocher and N. Ansell (Eds.), *Geographies of Global Issues: Change and Threat, vol. 8: Of the Series, Geographies of Children and Young People* (Ed. T. Skelton), Springer Reference, Singapore.

Freeman, C., Lingam, G.I., and Burnett, G. 2015. Children's changing urban lives: A comparative New Zealand-Pacific perspective. *Journal of Urban Design* 20(4): 507–525.

Freeman, C., Nairn, K., and Gollop, M. 2015. Disaster impact and recovery: What children and young people can tell us. *Kōtuitui, New Zealand Journal of Social Sciences Online* 10(2): 103–115.

Giacaman, R., Abu-Rmeileh, N.E.M., Husseini, A., Saab, H., and Boyce, W. 2007. Humiliation: The invisible trauma of war for Palestinian youth. *Public Health* 121: 563–571.

The Guardian. 2015. www.theguardian.com/environment/2015/jun/10/robin-wins-vote-uk-national-bird-britain

The Guardian. Nature is the best way to nurture pupils with special educational needs. www.theguardian.com/teacher-network/2016/may/01/nature-nurture-

Guimond, E., Robitaille, N., and Senécal, S. 2009. Aboriginal populations in Canadian cities: Why are they growing so fast? *Canadian* Winter: 11–19.

Gundersen, V., Skar, M., O'Brien, L.O., and Follo, G. 2016. Children and nearby nature: A nationwide parental survey from Norway. *Urban Forestry and Greening* 17: 116–125.

The Herald. 2010. www.nzherald.co.nz/nz/news/article.cfm?c_id=1&objectid=10675565 Thursday 23 September 2010.

Jenkins, K.J., Harte, H.M., Ririki, and Te, K. M. 2011. *Traditional Māori Parenting: An Historical Review of Literature of Traditional Māori Child Rearing Practices in Pre-European Times*, Te Kahui Mana Ririki, Children's Commission, Wellington.

Karsten, L. 1998. Growing up in Amsterdam: Differentiation and segregation in children's daily lives. *Urban Studies* 35(3): 565–582.

Karsten, L. 2015. Middle-class childhood and parenting culture in high-rise Hong Kong: On scheduled lives, the school trap and a new urban idyll. *Children's Geographies* 13(5): 556–570.

Keane, K. 2015. Ngā Manu: Birds. *Te Ara: The Encyclopaedia of New Zealand*. www.TeAra.govt.nz/en/nga-manu-birds/e.aspx accessed 29–10–2017.

Kinoshita, I., and Woolley, H. 2014. Children's play environment after a disaster: The great East Japan Earthquake. *Children* 2: 39–62.

Kostelny, K., and Garbarino, J. 1994. Coping with the consequences of living in danger: The case of Palestinian children and youth. *International Journal of Behavioural Development* 17(4): 595–611.

Kreutz, A. 2014. Lack of child-environment congruence inn Cherbourg, Australia: Obstacles to well-being in an indigenous community. *Children, Youth and Environments* 24(1): 53–81.

Malone, K. 2016. Theorizing a child: Dog encounter in the slums of La Paz using post-humanistic approaches in order to disrupt universalisms in current 'child in nature' debates. *Children's Geographies* 14(4): 390–407.

Māori Affairs. 2012. *Select Committee Submission: Inquiry into the Determinants of Wellbeing for Māori Children*, NZ, Govt, Wellington.

Mikkelsen, M.R., and Christensen, P. 2009. Is children's independent mobility really independent? A study of children's mobility combining ethnography and GPS/mobile phone technologies. *Mobilities* 4(1): 37–58.

Miller, J. 2003. Daoism and nature. In H. Selin (Ed.), *Nature across Cultures: Views of Nature and the Environment in Non-Western Cultures*, Kluwer Academic Publishers, pp. 185–209.

Nilon, C. 2014. The urban system: Social drivers. In R.A. McCleery, C.E. Moorman, and N. Peterson (Eds.), *Conservation: Theory and Practice*, Springer, New York. New South Wales Service for the Treatment and Rehabilitation of Torture and Trauma Survivors, Australia. www.startts.org.au/about-us/ accessed December 2016.

Nilsen, R.D. 2008. Children in nature: Cultural ideas and social practices in Norway. In A. James and A.L. James (Eds.), *European Childhoods: Cultures, Politics and Childhoods in Europe*, Palgrave McMillan, London.

Parkes, G. 2003. Winds, waters, and earth energies: Fengshui and sense of place. In H. Selin (Ed.), *Nature across Cultures: Views of Nature and the Environment in Non-Western Cultures*, Kluwer Academic Publishers, pp. 185–209.

Richter, B. 2001. *Mandela's Children*, Routledge, London.

Saldivar-tanaka, L., and Kransky, M.E. 2004. Culturing community development, neighbourhood open space, and civic agriculture: The case of Latino community gardens in New York city. *Agriculture and Human Values* 21: 399–412.

Save the Children. 2011. *Children Displaced by Japan Earthquake Face Shortages*. www.savethechildren.org.uk/2011-03/children-displaced-japan-earthquake

Selin, H. 2003. Introduction. In H. Selin and A. Kalland (Eds.), *Nature across Cultures: Views of Nature and the Environment in Non-Western Cultures*, Springer Science+Business Media, Dordrecht.

Sponsel, L.E., and Natadecha-Sponsel, P. 2003. Buddhist views of nature and the environment. In H. Selin (Ed.), *Nature across Cultures: Views of Nature and the Environment in Non-Western Cultures*, Kluwer Academic Publishers.

Swart Kruger, J., and Chawla, L. 2002. 'We know something someone doesn't know': Children speak out on local conditions in Johannesburg. *Environment and Urbanization* 14(2): 85–96.

Taylor, A. 2013. *Reconfiguring the Natures of Childhood*, Routledge, Oxford.

Ulster wildlife. n.d. *Bog Meadows*. www.ulsterwildlife.org/reserves/bog-meadows

UNICEF. 2013. *The State of the World's Children 2013*. UNICEF, New York.

United Nations Office for the Coordination of Humanitarian Affairs. 2012. Occupied Palestinian Territory Olive Harvest Factsheet. October 2012.

Von Benzon, N. 2011. Who's afraid of the big bad woods? Fear and learning disabled children's access to local nature. *Local Environment* 16(10): 1021–1040.

Woodgate, R.L., Edwards, M., and Ripat, J. 2012. How families of children with complex care needs participate in everyday life. *Social Science and Medicine* 75(10): 1912–1920.

Woolley, H., Armitage, M., Bishop, J., Curti, M., and Ginsborg, J. 2006. Going outside together: Good practice with respect to the inclusion of disabled children in primary school playgrounds. *Children's Geographies* 4(3): 303–318.

Wynd, D. 2013. *Child Abuse: What Role Does Poverty Play?*, Child Poverty Action Group, Auckland.

9 Mediated nature

Different forms of contact

There are three principal ways that children access nature: direct, indirect and vicarious contact (Kellert, 2002). Direct contact is the primary focus of this book and is when children made physical contact with nature through a real-world experience, such as climbing a tree, watching a wild animal or playing in a forest. Indirect is when children experience nature but in a controlled or mediated way, such as watching fish in a tank, going to a museum, pony riding or growing a pot plant. Vicarious nature is that which takes place through a medium, such as artworks, books, films or computer games or through other devices that use or present nature but there is no actual contact with real nature. In this chapter, we explore how children's contact with and understanding of nature can be both enhanced and frustrated by different forms of media. There have been strong arguments put forward, especially by the 'nature deficit disorder' adherents, arguing that children are becoming disconnected from nature with the primary protagonist in this being electronic media, traditionally television but increasingly digital media and games. One of the best known studies illustrating this reported disconnect was in *Science* magazine (Balmford et al., 2002), which found that 8-year-old children were more adept at naming Pokémon characters than common UK wildlife species (78% success rate compared to 53%).

We explore whether there has been a switch from 'biophilia' to 'videophilia'[1] amongst children. We also examine the role of digital media and other less-controversial, more-traditional media, such as stories, books, films and wildlife documentaries, in children's urban-nature experiences. In doing so, we investigate the ways nature is presented and packaged for children and how this potentially influences children's nature relationships. We explore the comparative lack of urban nature in mediated forms and the potential impact of an overriding dominant presentation of nature as pristine, rural and usually populated by large, cuddly or fierce creatures. It is a nature that may bear little resemblance to the everyday nature present in city children's lives. We argue for a more realistic and accessible 'mediated' nature. We also recognise the role that mediated nature can contribute to providing fun, engaging and sometimes instructive ways for children to start thinking about nature. Mediated nature is expanding in its forms but has a long history of getting children to relate to and think about nature. For example, favourite rhymes for our own children were 'Ooey gooey', which uses humour to draw attention to one of nature's less illustrious and in this case more unfortunate creatures, and one about the sponge.

Ode to a Worm (Anonymous[2])

Ooey Gooey was a worm,
A wondrous worm was he.
He sat upon a railroad track,
A train he did not see.
The train came roaring round the bend,
The driver gave a squeal,
The guard got out his pocketknife,
And scraped him off the wheel.
Ooey Gooey!

The Sponge (A. P. Herbert)

The sponge is not as you suppose a funny kind of weed,
It lives beneath the deep blue sea,
an animal like you and me,
though not so good a breed.
And when the sponges go to sleep, the fearless diver dives.
He prongs them with a cruel prong,
And what I think is rather wrong,
He also prongs their wives.
I know you'd rather not believe such dreadful things are done,
Alas, alas, it is the case,
And every time you wash your face, you use a skeleton.
So that is why I seldom wash, however black I am,
And use a flannel if I must,
Though even that, to be quite just, was once a little lamb.

Mediated nature: problem, asset or both?

We address the question here of whether children can connect to nature through media, and if so, do the media (books, computers, art, games and so on) replace nature or act as a vehicle that encourages and facilitates nature contact? There is a new term that is frequently used when talking about children and nature, 'videophilia', defined as 'the new human tendency to focus on sedentary activities involving electronic media' (Zaradich and Pergams, 2007, p. 130). In pre-computer times, there were concerns expressed about 'bibliophilia'; i.e., children who preferred to sit and read books rather than going outside. There is clear evidence that children's outdoor time is diminishing in many parts of the world. What is less clear is whether this is because of the greater attraction of electronic media, notably games, or because children are less able to go outside due to parental and other restrictions such that they now seek alternative entertainment through indoor, media-based activities? There is common reference in media and in conversations around environmental education to a dichotomous choice for children between outdoors/nature and indoors/electronic media, but reality is rarely so simple. In this book, we have identified a raft of factors affecting nature access and experiences. These include busier lives for parents and children, leaving less free time, enhanced pressure to achieve in educational, cultural, sport and other activities, perceptions of lower environmental safety, mainly from strangers and increased traffic and densification of cities, leaving less green space and smaller families, meaning fewer children available to play with outdoors.

One aspect which needs to be explored is media as a proxy for nature connection. This is not to argue that media can replace nature but to recognise that media can also act as an incentive for children to connect with nature. It can also function as preparation for natural experiences and as a way of maintaining connection when direct contact is not possible. The UK Wildlife Trust has webcams stationed at peregrine falcon nesting sites at Derby cathedral, in Nottingham and Aylesbury city centres (UK Wildlife Trust), and the Conserve Wildlife Foundation of New Jersey (Conserve Wildlife Foundation of New Jersey) has a camera directed at peregrines nesting on a skyscraper in their city. Further afield, children can access live webcams of Loch Ness, which show grazing sheep but potentially could provide a glimpse of the mythical Loch Ness monster (Takala, 2017). People can access webcams at many wildlife destinations, such as the webcams at African waterholes (Explore.org, 2017). A globalising and increasingly online world means that nature experiences previously reserved for the few who could go on an African safari or to the top of Derby Cathedral or travel to Loch Ness are now available to a much larger group of children and are available irrespective to some degree of socio-economic status and location. Children's lives and the environments they live in are changing rapidly. As such, we need to know how to harness these changes and the changing media to enable children's nature understanding, experience and empathy rather than allowing the media to dictate modes of engagement and outcomes. Next, we examine the concept of eco-literacy, i.e., how children become informed about and engage with nature through key media, namely storytelling, books, poetry, films, interactive media (such as digital games) and mobile digital technologies.

Stories and poems

Storytelling is the oldest form of communication with children about nature. Children need stories to find out about the world and for knowledge to be passed on. Stories and the language they use are powerful ways to nurture 'biophilia' (Bai et al., 2010). Bai et al. reason, "[I]t makes sense to assume that all children in their embodied totality start out as potential lovers of nature, full of curiosity and wonder' (p. 358). Hence, evocative images, such as the full moon, dark forests, silvery seas, storm clouds and sunlit meadows are commonly found in stories. Storytelling about nature performs different functions – fun, engagement, distraction, passing on of knowledge, preparation for later life, enhancing understanding, teaching respect and caring and communicating important life affirming and safety values. Many of the folk tales that relay messages about nature have long and continuing historical roots. Little Red Riding Hood is still a popular children's story, though latterly more usually read than told orally. It tells of the dangers of forests and wolves. It is usually best known as one of the Brothers Grimm Fairy Tales, although variations of the story date back to the 11th century (Tehrani, 2013) and even earlier. Another still popular story, also with a wolf at its centre, is the Aesop fable "The Boy who Cried Wolf", with parents still explaining the idiomatic expression of 'don't cry wolf' through the medium of the story. There is some concern that oral storytelling is increasingly being usurped by stories that are read, rather than transmitted through oral retelling or that are transmitted through visual or digital media.

Poetry and rhymes in particular have been the focus of oral learning for many years and are still common with very young children. The following examples selected from the UK and internationally (Table 9.1) are commonly used with children and include mention of animals such as birds and insects, plants, the weather and the night sky. Whilst all possess rhythm, are short and designed to catch the child's attention and are easy to mimic, they all reference nature and in some cases a sense of wonder.

For First Nations children and children who live in cultures that still have rich oral histories, storytelling is an especially vital conduit of knowledge.

> Historically and today, First Nations people share important knowledge, culture and traditional lessons through the telling of stories. It is through the telling of stories and legends that First Nations people preserve what is most important to them – language, traditions, culture, and identity. Stories are used to provide a sociocultural and historical account of the community knowledge from elders to youth, ensuring its survival with new generations.
>
> (Fixico, 2003 in McKeough et al., 2008, p. 5)

Children learn for themselves early on the power of narration in their own stories and frequently invoke images of or experiences in nature, with nature being especially important in fantasy stories. Gómez and Maker (2011) examined 87 stories from children (generally aged 8) living in urban areas of the US and found children tended in their stories to frequently write about settings that involved animals and nature. Native American and Hispanic children tended to include nature as a central topic, especially in their fantasy stories, thus building on traditional storytelling, where creatures such as the coyote or animals with magical or healing power frequently appear (Gómez and Maker, 2011). Nature-based knowledge is often a critical component in stories both told to and told by children, but there seems to be little that relates to nature and the reality of life in an urban setting, although many simple stories for early children are location free. Traditional tales still tend to reflect historic nature in rural or village settings, as was dominant at the time of their creation, e.g., the Ugly Duckling, Hansel and Gretel and Monkey-Crab Battle (Japan) with similar stories found in most cultures. However, stories as a primary source of cultural knowledge are under pressure from more modern media, which detract attention from orally transmitted knowledge. In storytelling, books are a vitally important source of knowledge.

Table 9.1 Rhymes with nature features commonly known by children from around the world

English nursery rhymes			
Ladybird, ladybird, fly away home. Your house is on fire and your children are gone, All except one, And her name is Ann, And she hid under the baking pan.	Mary, Mary, quite contrary, how does your garden grow? With silver bells, And cockle shells, And pretty maids all in a row.	Incey Wincey Spider climbed up the spout, down came the rain And washed the spider out, Out came the sun And dried up all the rain, and Incey Wincey Spider climbed up the spout again.	Twinkle, twinkle, little star. How I wonder what you are, Up above the world so high, like a diamond in the sky.
From around the world, home language and English translation*			
Muti Muhande* **A Tree Planted** **(Kikuyu)(English)**	**Imvula** **It's Raining** **(Xhosa)(English)**	**Der Hahn ist tot** **The Rooster Is Dead** **(German) (English)**	**Cuddly Koalas** **(Australia)**
Muti muhande rui-ini One i mathangu maguo Magithaka na ruhuho Na ithui nituthake na ruhuho. (x4) A tree planted on the river bank, look at its leaves playing with the wind. Even us let's play with the wind.	Imvula, imvula, Chapha (1), chapha, chapha Chapha, chapha, chapha Imanz' impahla (2) yam. Imanz' impahla yam. Gqum (3), Gqum, Liyaduduma Gqum, Gqum, Liyaduduma Imanz' impahla yam. Imanz' impahla yam. It's raining, it's raining, Chapha, chapha, chapha. Chapha, chapha, chapha, My clothes are getting wet. My clothes are getting wet. Gqum, Gqum, It's thundering! Gqum, Gqum, It's thundering! My clothes are getting wet. My clothes are getting wet.	Der Hahn ist tot, der Hahn ist tot, Der Hahn ist tot, der Hahn ist tot, Er kann nicht mehr kräh'n, kokodi, kokoda, Er kann nicht mehr kräh'n, kokodi, kokoda. Kokokokokokokokodi, kokoda The rooster's dead, the rooster's dead. The rooster's dead, the rooster's dead. He'll no longer cry kokodi, kokoda. He'll no longer cry kokodi, kokoda. Kokokokokokokokodi, kokoda.	Cuddly koalas, Cuddly koalas. Possums, too. Possums, too. Wallabies and wombats. Wallabies and wombats. Kangaroos. Kangaroos.

* www.mamalisa.com/?t=es&p=1716

Books

Books transmit knowledge as well as provide for literary and other facets of children's development. Given their significance in the lives of most children, particularly in the global north and increasingly in the global south, it is useful to see how cities and nature are represented in children's books. Popular post-war stories, such as those written by UK author Enid Blyton, show children having free access to the countryside for exploring and adventures. At around the same time in Germany, the Bullerby novels of Astrid Lindgren show the Bullerby children, like their compatriots in Enid Blyton's Famous Five, as similarly carefree children having adventures in the countryside. The term *Bullerby* has since been used in the work of Broberg et al. (2013) and Kyttä (2004) to examine children's outdoor play comparable to that represented in the novels of Lindgren's children and their lives in a Swedish village. Bullerby is used to evoke a notion of a childhood full of environmental affordances[3] and children's independent experiences. However, one of Lindgren's other famous characters, Pippi Longstocking, lives in a town (albeit a small one) and reveals a similar sense of adventure. There is no current series that similarly evokes the sense of adventure and freedom for children living in cities, as if to suggest that for city children, adventure in nature is not possible. Moreover, at a more general level, a study undertaken by Williams et al. (2012) of 8,067 illustrations from 296 award-winning children's books dating between 1938 and 2008 suggested nature itself was becoming less evident in children's stories. Their findings can be summarised as follows:

- Caldecott[4] winners and honour books have given less attention in recent years to the natural world and more attention to build environments.
- Just as images of nature have declined, so have the images of animals.
- There is also less human interaction with animals than in the past. Scenes of interaction with wild animals have never been common but have become even less so in recent years.
- Environmentalism has not taken a prominent place among the 296 books examined.

In light of this study, we undertook our own analysis of 50 children's books to see how nature was present in books. The method we used was simple and does not pretend to have scientific rigour. We used the first 50 books selected through Pinterest and Amazon (the largest online bookseller), using nature and children as search terms, to identify what trends are present. The list includes some children's favourites such as *The Very Hungry Caterpillar* (Eric Carle), *The Gruffalo* (Julia Donaldson), *The Lorax* (Dr Seuss) and *Giraffes Can't Dance* (Giles Andreae, Guy Parker-Rees). Only three books showed realistic nature portrayals: *Flotsam* (David Wiesner), *Swirl by Swirl* (Joyce Sidman and Beth Krommes) and *An Egg Is Quiet* (Dianna Aston and Sylvia Long). All three celebrated the beauty and intrigue of nature. In 34, the main character was an animal, with the next commonest main character being a child. Animals thus are very likely to be centre stage. This stage though was rarely an urban one, and if it was focused on a settlement, it was invariably a village or farm and had a European setting. Only four had town or city settings.

Anthropomorphising animals in children's stories is common and has a long history, from Peter Rabbit (Beatrix Potter) to more recent forms such as *Charlotte's Web* (E. B. White), which features a pig named Wilbur and a spider named Charlotte. Giving animals human traits enables children to empathise with animals and see the world from a different perspective, albeit a dominantly human one, despite the animal character. Of the 50 books, some 27 included an anthropomorphic animal and another 15 showed some anthropomorphic traits, so clearly the genre is alive and well. Some stories provided more realism, dealing with animals behaving primarily as animals, such as *Watership Down* (Richard Adams) and several of Beatrix Potter's tales, and of our 50, about eight engaged with natural history in some form. One surprising exception from this list is *We're All Going on a Bear Hunt* (Michael Rosen and Helen Oxenbury). In the bear hunt, the family and the children enter a natural world as they travel through the landscape (grass, river, mud, forest and snow) on their quest to find a bear (Figure 9.1). In *Where the Wild Things Are* (Maurice Sendak), Max encounters an imaginary place populated by wild imaginary creatures. Both stories engage children with the concept of the possibilities present in nature and feelings that are both positive and scary

Figure 9.1 Even very young children are drawn into the 'We're All Going on a Bear Hunt' adventure. In this photo, they cross the meadow going through the 'swishy swashy' grass.

associated with nature encounters. Largely absent from the 50 books though are books that show nature in urban or city settings. Where animals do feature in children's books, they tend to be domesticated animals rather than wild animals and the animals frequently exhibit identifiably human rather than animal attributes.

The centrality of animals in children's literature led Flynn in his assessment of animal stories to suggest, "[A]nimal stories have become so integral to children's literature that the form is nearly unthinkable without them" (Flynn, 2004, p. 48). In this though, there is a puzzling contradiction as identified by Fawcett in her study of trends in children's wild animal stories. She observes, "[I]t is perplexing, though that as more live wild animals disappear we continue to inundate children with animal symbolism. We give them stuffed animals . . . endless animal allegories, stories, movies and cartoons . . ." (Flynn, 2004, p. 133). If we look into the presentation of nature in children's stories, several trends emerge in addition to the dominance of anthropomorphised animals. In many cases, there is little attention to accuracy or realism in the presentation of nature: natural features and animals frequently appear out of their geographic context, and the mix of animals is 'wrong'. For instance, a popular story at our children's library is *The Very Cranky Bear* (Nick Bland) about a bear who has friends who are a sheep, lion, zebra and moose. Or *Cuddley Dudley* (Jez Alborough), a story about a penguin that clearly takes place in the Arctic, despite the fact that penguins are restricted to the Southern Hemisphere. In contrast, the animals in *Giraffes Can't Dance* are all at least African rather than multi-continental. Rarely too is the cycle of life in nature accurately rendered outside of books with an explicit ecological or environmental message. As the saying goes, 'nature red in tooth and claw' but not in children's books or films, except perhaps for National Geographic and David Attenborough's animal films, which can be very 'bloody'. This is interesting in light of children's own experiences, where in their daily lives, many children will encounter birds, hedgehogs or other animals killed on roads or washed up on beaches, and such experiences can provoke considerable interest (Figure 9.2). An exception is the well-loved book *The Lorax* (Dr Seuss), in that it follows the destruction of nature through over-exploitation, the loss and pollution of the habitats of the swomee-swans, the brown bar-ba-loots and the hummingfish and the ultimate extirpation of the beautiful truffula trees. The tale is saturated with regret for what has happened but with a positive twist in that it can be rectified if the right person comes along and cares enough to make it better.

What about nature and the city and how this is presented in children's books? There is a wide continuum from nature-rich cities to cities being presented as a natural wasteland – an urban dystopia from which nature has been banished. Another story that extends across the whole continuum is about Rose, an adventurous girl who travels in a teapot collecting seeds, which she then uses to transform a patch of earth in the city into a glorious colourful city garden that everyone uses (*Rose's Garden*, Peter H. Reynolds). In 2015, an award-winning book, *Footpath Flowers* (Sydney Smith) shows a little girl going on a city walk through a black-and-white denatured city, finding flowers in the footpath that lead her to greater awareness of city flowers and nature; her walk ends in a book that is effervescent with colourful flowers. Stories such as *Footpath Flowers* that show the nature present in everyday

Figure 9.2 This girl was very impressed by the remains of a jellyfish that she found on a city beach in Newcastle, Australia.

city environments are rare, and there is little evidence of the current movement of greening of cities present in children's literature. However, despite the paucity of nature-based city books, there is one particularly and astoundingly beautiful book *Uno's Garden* (Graeme Base), which does offer hope, and indeed, hope is its main concern. In many ways, the book echoes the theme of *The Lorax*, where a biodiverse natural landscape becomes spoiled and the exploiters come to regret its loss. But the activities of Uno's family mean that species are saved and in turn multiply such that a harmonious city-nature relationship develops. While both *The Lorax* and *Uno's Garden* offer hope, one could ask why the need to place this message in a 'fantasy' setting? Is it because such achievements are seen as unrealistic or impossible in real settings in real cities? This type of oral-based storybook learning seems to be coming under pressure from what Arleo calls the "assaults of modern life and technology" (Arleo, 2004), leaving less time for oral and possibly written books as more animated visual media take precedence.

Films and television

Next we look at films and see how nature is presented, in particular city nature. Animals and nature are popular themes in children's films. Disney uses animals as their central characters in many films – *The Lion King*, *Bambi*, *The Jungle*

Book, *The Fox and the Hound*, *Zootopia* and numerous cat and dog stories to name but a few. Pixar similarly produces animal-based films, including *Finding Nemo*, *A Bug's Life*, *Ratatouille* and *Finding Dory*. Another major player is DreamWorks, whose films include *Over the Hedge*, *Madagascar*, *Happy Feet*, *Antz*, *Spirit*, *Ice Age*, *Kung Fu Panda* and *Shark Tale*. Booker (2010) argues that Disney has a history of consistently championing nature, often in opposition to human abuse of nature, but admits that the nature presented is of a peculiar type. There are other non-animated films, but given the scope and influence of the animated films for children, the nature-based messages they portray are significant.

The portrayal of animals and natural relationships in these films poses many challenging questions. Does it matter, for example, that in *The Lion King*, Simba is best friends with a warthog (normally a prey species) and seems to eat nothing more than grubs? The cycle of life, which is a key theme in the film, in actuality involves more than falling in love and having a cub as Simba does. A central component of the cycle of life really means the relationship between predator and prey, where Simba, in the real world, would be dependent on food species, such as the antelope that in the film are shown as lauding his coming of age as king, not fleeing from their hunters. It also means understanding animal relationships such that hyenas are not just evil, scheming creatures as in the film but fulfil an important and positive role in the food chain. In *Happy Feet*, Mumbles, the main character, never completes his moult from chick to juvenile or adult plumage. In reality, if he had entered the water still with down feathers rather than full adult plumage, he would have died very quickly from heat loss. Similarly, does it matter that the premise in another film, *Shark Tale*, is that the good shark is a vegetarian and therefore not like his 'killer' family? It is clearly a false premise given his species. The categorising of species into good and bad is a recurrent theme in the films. Snakes (*The Jungle Book*) and hyenas (*The Lion King*) are invariably bad, as are sharks (*Jaws*), whereas deer (e.g., Bambi) are good and innocent, pandas are endearing (*Kung Fu Panda*), and mice are often cute and clever (e.g., *Stuart Little*, *The Great Mouse Detective* and *Tom and Jerry*).

The Smithsonian (reputedly the world's largest museum and research complex, Washington, DC) identified their top ten children's[5] films with a green or nature theme as follows:

1. *FernGully: The Last Rainforest*
2. *WALL-E*
3. *Bambi*
4. *Over the Hedge*
5. *Hoot*
6. *Star Trek IV: The Voyage Home*
7. *Free Willy*
8. *Disney Nature's Oceans*
9. *Avatar*
10. *Happy Feet*

In this list of the ten best films, what is especially interesting for this book is that the only one set in a determinedly city setting is *WALL-E*, which is dystopian in its presentation of the city. In *Over the Hedge*, the problem is the encroaching town, a realistic message given the habitat destruction afforded by uncontrolled urban sprawl. Again, a positive city-nature connection is lacking in this list of ten. If we move beyond English-language films, Studio Ghibli's Japanese studios produced an animated comedy-drama fantasy film *Pom Poko*. This film features tanuki (based on Japanese raccoon dogs) who become disaffected through city encroachment. The film is environmentally oriented, asking its audience to consider the impacts of human invasion of wildlife spaces. The portrayal is darker and, despite its fantasy overlays, more realistic than movies such as *Over the Hedge*. For the tanuki, there is no happy ever after; some survive by melding into human society, some are exiled and some survive in parks, but overall, there is no positive outcome for the tanuki. The messages contained in these depictions of the impact of urban encroachment on wildlife are thus universal and in this sense draw children's attention to the need to recognise that cities do have an impact on other species and that they do have a responsibility to provide for nature and its displaced species.

Visual media through films and TV can also act to encourage direct contact. A popular TV series directed at preschool children is *Dora the Explorer*, where children are encouraged to go out and explore, with the episodes often focusing on natural features, such as the beach, a storm, a star, a mountain, a crab, a forest and sundry animals. The series has been translated into around 30 languages, including Mandarin, Polish, Thai, Malayalam and Māori, and books, videos and computer games have supplemented the TV series. It is a potentially very powerful medium for guiding young children's thinking and understanding around nature, while simultaneously encouraging them to go out and explore it.

Documentaries

A major influence on children's views about nature are documentary-type films. In this genre, a perpetual favourite for both children and adults is David Attenborough, broadcaster and naturalist. He is best known for his natural history films, most notably the Life series, beginning with *Life on Earth* (1979) and including *The Life of Birds*, *The Life of Mammals* and *Planet Earth*. With stunning cinematography and capturing superbly the intimate lives of species, David's Attenborough's narration draws the watcher in. In terms of engaging children, Attenborough is a sure winner. However, again, the emphasis is invariably on the rare, the spectacular and a perpetuation of the belief that nature is primarily 'over there'. Only in 2016 did Attenborough make a film focusing on the city, in this case Singapore, in the film, *Wild City*.

Films in the form of documentaries have also been used to promote children's relationships with the outdoors and to highlight children's nature-based relationships. There are a growing number of films designed to expose what the filmmakers see as children's lack of outdoor connection. A recent example is from *Project Wild Thing*. This documentary takes its motivation from the need to address the perceived rise of 'nature deficit disorder' and the other negative repercussions associated with a lack of outdoor and wild play. The short film, which uses the image of a young girl licking a frog in its advertising material, states,

> Today's young people stay indoors, unlike their parents and grandparents, who as children loved to roam free. A new generation is addicted to iPad and consoles, getting obese and insidiously depressed in the process . . .
>
> (The Guardian, 2013)

The documentary shows city children in a particularly concerning way, as being denatured and disconnected. There are other similar documentaries, such as Miranda Andersons's[6] short film *The Child in Nature*, which features an interview with Richard Louv and opens with a clip showing a boy playing a computer car racing game with the words 'Technology is taking over'. Miranda is herself a middle school child so offers an interesting youth perspective. With a similar message but less obviously proselytising is a Scandinavian film *NaturePlay: Take Childhood Back* (Stilling and Stilling, 2016). This documentary celebrates children's engagement with nature, showing them experiencing forests and fiords and genuinely wild landscapes. Though stunningly beautiful and inspiring, the notion of children going into nature again predominates. It opens with the words: "The data beast prowls in US schools" and "The most endangered species in the wild today: our children". The film portrays the Danish tradition of taking children from an early age into nature as part of their normal education. It recognises the value of forest school education, which is common in Denmark. While differing in approach, these films nonetheless show some commonalities. First, they all see children as endangered and in danger of becoming divorced from nature, and they target the computer as an important negative influence. Second, the images are predominantly of children being taken out into 'wild' nature in country settings. There seems to be little acknowledgement of the wild in cities and around children's homes. While the films' images of children running and playing in 'pristine' woodland and other similar environments are powerful, the message of nature as 'over there' is perpetuated. A documentary celebrating children's engagement with the wild in their backyards and focusing on city environments is significant by its absence and could provide a valuable alternative narrative.

Films, books, poetry, songs and art all provide ways for children to think about nature, to explore and understand it, but the context needs to better reflect the reality of most children's existence, namely an urban one. They need to demonstrate that children can realistically connect with nearby nature, not just the 'out there' nature they experience through the media. The one media that has not yet been referred to is that of technology, computers and mobile technology. These have been the focus of much debate about their potential to majorly impact on children's relationships with the outdoors.

Interactive media – computers and games

All of the three documentaries referred to in the previous section make direct links between the lack of nature contact and computers. References in the research literature make this connection seem incontrovertible as the following selected extract indicates:

> After 50 years of steady increase, per capita visits to national parks have declined since 1988. This decline is coincident with the rise in electronic media . . . We may be seeing evidence of a fundamental shift away from people's appreciation of nature to videophilia.
>
> (Pergams and Zaradich, 2006, p. 387)

A review of the clinical and psychological effects of excessive screen time on children reports,

> The explosion of new technological devices over the last few years has led to electronic media becoming an integral part of life. As such, it may be presumed that adherence to screen time guidelines has further declined. Exposure to screens tends to start from very early infancy. Studies estimate that children spend a substantial proportion of their daily waking hours on screen-based activities, about 8 h [hours] in many cases.
>
> (Domingues-Montanari, 2017, p. 333)

There is undeniable evidence that children's use of digital media is increasing. Childwise UK (a specialist organisation researching children and young people) undertakes an annual study of children and young people across the UK and for 2016 reported that

> Children age 5–16 now use the internet for an average of three hours a day and watch TV for 2.1 hours. It is worth noting that simply counting hours spent on devices can be problematic. Children multi-task and often use more than one device at once.
>
> (Childwise, 2016)

While there is good cause for concern about the impacts of excessive screen use (Daily Mail, 2015), given that electronic media are here to stay and their use amongst children globally is likely to increase, it is important that their role in children's lives be considered. This is not simply to assess their negative impacts on children's behaviour but also to reference the ways that electronic media can potentially be harnessed to assist children's connections with the natural world (Figure 9.3).

Similar to the media already profiled (books, films and so on), computers can provide vicarious nature experiences through which positive environmental behaviours can be supported and developed and provide for positive behavioural change when outdoor opportunities are lacking (Kil, 2016). To be beneficial,

Figure 9.3 Computers and computer animation are engaging for children and can be used to encourage development of interest in nature. This child is watching an animated version of *The Very Hungry Caterpillar* story.

the virtual experience needs to be realistic and carefully thought through. The messages in video games that children use are not always positive, as in the 'shoot and kill games' or where nature is presented in inappropriately anthropomorphised or misleading ways. Conversely, two games that make positive attempts to enable children to engage nature online positively are *Wilderquest*[7] and *Rainforest Survival Challenge*[8], both profiled in an engagingly titled paper "Gaming conservation: Nature 2.0 confronts nature-deficit disorder". In his paper, Fletcher (2017) concludes that these games present aestheticized, hyperreal images of nature, but do, nonetheless, also directly address conservation issues. However, as with all such games, there is the ever present danger that as technology advances and virtual reality becomes more accessible, the games will be seen as an alternative and even a better experience than the real thing. Mosquitos in games do not bite, and you do not get cold and wet. Fletcher argues that though this anxiety exists, such games are more likely to enhance rather than diminish concern about nature. However, as with the earlier analysis of books and films, representations of positive nature in computer games seem lacking. For older children and adults, the dominant image is too often one of a dystopian future in a denuded city.

There are a number of city building games. The best known is perhaps the SimCity series, but there are others; the Tropico series, Banished, Imagine Earth, Cities: Skylines, Lethis – Path of Progress', and some such as 'Project Aura' and 'Startopia' that are set in a post-apocalyptic future. The earth-based city-building games do include grass, trees and as necessary farm animals and natural features such as rivers and seas. The city settings may have a green appearance, but there is little evidence of any substantive natural thinking with no concerted effort towards creating biodiverse living environments. Digital games set in the city provide plenty of potential for embracing biodiverse environments. Developments in graphic capability mean that it should be possible to make more realistic urban landscapes, certainly more so than the somewhat simplistic, stock images of green lawn and trees that currently predominate.

So far, the emphasis has been on games usually played on a static computer. Games, especially those designed for mobile phone applications, can also be great motivators for getting children outdoors. Geocache was a precursor for some of these games where GPS coordinates lead children to a hidden cache somewhere outdoors (Figure 9.4). Released in 2016, *Pokémon Go* is a free-to-play, location-based augmented reality game that replicates aspects of real-world wildlife watching and natural history by requiring players to locate Pokémon-related features at different locations in the real world. The game has encouraged huge numbers of people to go outdoors and increased their time spent there; its release in New Zealand and elsewhere was followed by the common sight of children out and about, using their devices to locate Pokémon sites (Figure 9.5). The motivation of the designer of the first Pokémon game, Tajiri Satoshi, was to replicate his own

Figure 9.4 Various media, GPS, Google can be used to help children go out into nature to find geocache sites, such as the one hidden in the scene shown in this Google photo of Mount Cargill, Dunedin.

Figure 9.5 Two boys playing *Pokémon GO*, in Wanaka, New Zealand, against the stunning backdrop of Mount Aspiring National Park.

childhood experiences of watching birds and collecting insects and crayfish, but in a pre-industrial playworld, because he felt that rapidly growing urban areas offered limited opportunities for urban children to connect with nature (Allison, 2003). Dorward et al. (2017) explore the conservation implications of *Pokémon Go* and refer to increased motivation to visit natural sites and exposure to basic natural history concepts such as species habitat preferences and variations in abundance. Moreover, there is evidence that people discover non-virtual wildlife while playing *Pokémon Go*, leading to the Twitter hashtag #Pokeblitz, which helps people to identify real species found and photographed while playing. Potential negative aspects include the concept of 'catching' creatures that are subsequently used to fight each other, thus reinforcing exploitative use of nature and the possibility that the search for fantasy creatures might distract players from the real creatures around them. Dorward et al. (2017) go on to suggest ways to modify *Pokémon Go* itself to enhance nature connection by, for example, making Pokémon biology and ecology more realistic or adding real species to the *Pokémon Go* universe. Another game *Ingress* is a location-based, augmented-reality mobile game that demands the player capture 'portals' at strategic outdoor places. Ideally, these types of games act as a spur for children to explore and engage with the environments around them.

Mobile applications for smartphones and tablets can also facilitate nature connection by turning just about anyone into a citizen scientist. Collection of data by laypeople, including children, has increased in popularity, greatly facilitated by the ease and convenience of mobile devices, so that there are currently hundreds of citizen scientist projects taking place across the world. Those with a focus on the natural world are incredibly diverse, from tracking marine debris, bugs in the backyard, invasive species and migrations of butterflies and birds to exploring and documenting wildlife in the world around us, e.g., Project Noah. Noah, which is backed by National Geographic, is an acronym for 'networked organisms and habitats' and is designed to help laboratories, environmental groups and various organizations to gather important data for research projects. The website *SciStarter* lists more than 600 active citizen science projects around the world, one-third of which are powered by mobile apps. In New Zealand, Nature Watch NZ (Naturewatch) describes itself as 'a place where you can share what you see in nature, set up citizen science and community-based monitoring projects, meet other nature watchers, and learn about New Zealand's natural history' (2017). The facility to set up a 'project' has been used by schools to monitor their own nature programmes. Many projects have an urban focus, such as the 'Celebrate Urban Birds' project run by the Cornell Laboratory of Ornithology and projects focused on backyard and nearby-neighbourhood diversity. As such, they are a very positive example of how online technology and mobile devices can facilitate nature connection in urban settings.

Going forward with media to engage with nature

Books have always been seen as a great source of information about nature, but so too has the more negative image of the bookworm stuck inside while more energetic children go on adventures outside. TV has similarly been seen as both supporter of nature connection, especially through advocates such as David Attenborough, Steve Irwin and others, and as competitor when children are portrayed as couch potatoes transfixed by the TV. The latest in this list of 'villains' is the computer. As we have seen, the reality is actually complex. Computers and phones provide opportunities to gain new knowledge and to share experiences through mobile phone applications. Never has it been so easy for children and young people to share their outdoor experiences. Studies have shown that vicarious environmental experiences can enhance environmental knowledge, caring and interest. Indeed, Kil (2016) suggests that what we need to focus attention on ways:

> to convert the videophilia tendency into a more pro-environmental trend by using it to transmit more knowledge about nature . . . [and] fuel additional interest in visiting natural areas to experience first-hand what they have read or seen in various forms of media.
>
> (p. 12)

Media in all its forms needs to be part of the ongoing nature story as it plays out in children's lives. New forms of media will carry on developing new applications that will be available on phones, and technology is unlikely to lose its appeal for children and young people. The challenge is to engage with developments in ways that can be used to positively connect children with nature. In the process though, long-established media, such as poems, can still be part of the natural equation. We end this chapter with a poem "Ode to a Goose", which is commonly taught to Chinese children and was penned during the Tang Dynasty by seven-year-old Luo Binwang (619–684), who went on to become a famous poet.[9]

Goose, goose, goose,
You bend your neck towards the sky and sing.
Your white feathers float on the emerald water,
Your red feet push the clear waves.

Notes

1 Biophilia – affiliation with nature; videophilia – connection to electronic media.
2 As with many well-known poems and sayings, there are variations on the wording, though the fate of the worm seems to be common.
3 An environmental feature or object that offers the potential for action, e.g., a tree for climbing, a lake for swimming, or a slope for rolling down.
4 The Randolph Caldecott Medal annually recognizes the preceding year's most distinguished American picture book for children.
5 They did concede that the definition of children's film may be a bit liberal given the inclusion of *Avatar* and Star Trek, www.smithsonianmag.com/science-nature/top-ten-kids-movies-with-a-green-theme-19409104/.
6 Miranda Anderson, *The Child in Nature*, www.childrenandnature.org/2013/02/04/the-child-in-nature-a-new-film-by-miranda-andersen-13/
7 Supported by the NSW National Parks and Wildlife Service.
8 Designed by Rainforest Alliance.
9 Translations of the poem in English differ slightly – source http://blog.tutorming.com/mandarin-kids/5-famous-poems-that-every-chinese-kid-knows.

References

Allison, A. 2003. Portable monsters and commodity Pokémon as Japan's new global power. *Postcolonial Studies* 6(3): 380–395.

Arleo, A. 2004. Children's rhymes and folklore. In P. Hunt (Ed.), *International Companion Encyclopaedia of Children's Literature*, Routledge, Oxon.

Bai, H., Elza, D., Kovacs, P., and Romanycia, S. 2010. Re-searching and re-storying the complex and complicated relationship of biophilia and bibliophilia. *Environmental Education Research* 16(3–4): 351–365.

Balmford, A., Clegg, L., Coulson, T., and Tay, J. 2002. Why conservationists should heed Pokémon. *Science* 295(5564): 2367b.

Booker, K.M. 2010. *Disney, Pixar, and the Hidden Messages of Children's Films*, Praeger, CA.

Broberg, A., Kyttä, M., and Fagerholm, N. 2013. Child-friendly urban structures: Bullerby revisited. *Journal of Environmental Psychology* 35: 110–120.

Childwise. 2016. *Childwise Monitor Report 2016*, Connected Kids, Childwise, Norwich, UK.

Conserve Wildlife Foundation of New Jersey. *Jersey City Falcon Cam*. www.conservewildlifenj.org/education/falconcam/ accessed 28–4–2017.

Daily Mail. 2015. The-simple-outdoor-pleasures-missed-today-s-computer-loving-children. www.dailymail.co.uk/news/article-3019573/The-simple-outdoor-pleasures-missed-today-s-computer-loving-children.html#ixzz4fW3HrWp7 accessed 29–4–2017.

Domingues-Montanari, S. 2017. Clinical and psychological effects of excessive screen time on children. *Journal of Paediatric Child Health*, 6, 53(4): 333–338.

Dorward, L.J., Mittermeier, J.C., Sandbrook, C., and Spooner, F. 2017. Pokemon go: Benefits, costs, and lessons for the conservation movement. *Conservation Letters*, January/February, 10(1): 160–165.

Explore.org. *African Safari Camera*. http://explore.org/live-cams/player/african-safari-camera accessed 28–4–2017.

Fletcher, R. 2017. Gaming conservation: Nature 2.0 confronts nature-deficit disorder. *Geoforum* 79: 153–162.

Flynn, S. 2004. Animal stories. In P. Hunt (Ed.), *International Companion Encyclopaedia of Children's Literature*, Routledge, Oxon.

Gómez, P.M., and Maker, J.C. 2011. What are the themes in young children's stories? An analysis of the content of children's written productions. *Zbornik Instituta za pedagoska istrazivanja, ipi* 43(1): 86–105.

The Guardian. 2013. Project wild thing. www.theGuardian.com/film/2013/oct/24/project-wild-thing

Kellert, S.R. 2002. Experiencing nature: Affective, cognitive, and evaluative development in children. In P.H. Kahn and S.R. Kellert (Eds.), *Children and Nature: Psychological, Sociocultural, and Evolutionary Investigations*, Massachusetts Institute of Technology Press, Cambridge, MA, pp. 117–151.

Kil, N. 2016. Effects of vicarious experiences of nature, environmental attitudes, and outdoor recreation benefits on support for increased funding allocations. *The Journal of Environmental Education* 47(3): 222–223.

Kyttä, M. 2004. The extent of children's independent mobility and the number of actualized affordances as criteria for child-friendly environments. *Journal of Environmental Psychology* 24: 179–198.

McKeough, A., Bird, S., Tourigny, E., Romaine, A., Graham, S., Ottmann, J., and Jeary, J. 2008. Storytelling as a foundation to literacy development for aboriginal children: Culturally and developmentally appropriate practices. *Canadian Psychology* 49(2): 148–154.

Naturewatch. About Naturewatch NZ. http://naturewatch.org.nz/pages/about-naturewatch accessed 11–11–2017.

Pergams, O.R.W., and Zaradich, P.A. 2006. Is love of nature in the US becoming love of electronic media? 16-year downtrend in national park visits explained by watching movies, playing video games, internet use, and oil prices. *Journal of Environmental Management* 80: 387–393.

Stilling, A., and Stilling, D. 2016. *NaturePlay: Take Childhood Back*, Directed and Produced by Aimee Stilling Along with Co-Director and Cinematographer Daniel Stilling.

Takala, M. 2017. *Nessie On the Net Loch Ness Live Stream*. www.lochness.co.uk/livecam accessed 28–4–2017.

Tehrani, J.J. 2013. The phylogeny of little red riding hood. *PLoS ONE* 8(11): e78871.

UK Wildlife Trust. *Nottingham: Peregrines, Aylesbury Peregrine Falcons*. www.wildlifetrusts.org/webcams accessed 28–4–2017.

Williams, J.A., Podeschi, C., Palmer, N., Schwadel, P., and Meyler, D. 2012. The human-environment dialog in award-winning children's picture books. *Sociological Inquiry* 82(1): 145–159.

Zaradich, P.A., and Pergams, O.W. 2007. Videophilia: Implications for childhood development and conservation. *The Journal of Developmental Processes* 2(1): 130–144.

Part IV

Creating connections, making it happen – biodiverse environments for all children

10 Redressing the balance in challenging urban environments

Bringing nature back to where children live

Challenging environments

While arguably there is nature present in all environments, even the most built, connecting with nature can be particularly challenging in dense urban environments where only the most robust and adaptable species can thrive. In this chapter, we address environments where nature is least present and children can find connecting with nature most problematic because of a lack of nature or because of societal challenges. We explore environments where the built form creates conditions that are currently antagonistic to wildlife or where the green space present under-provides for biodiversity; i.e., inner cities, very high-density urban residential areas, high-rise tower blocks, roads and commercially dominated environments and, in some cases, school environments. We discuss the issue with reference to children living in challenging environments, which may be inner-city, high-density but may also be deprived estates on the edge of cities. High-density and nature deprivation is not limited to the inner city. In particular, we recognise the presence of societal factors associated with deprivation and poverty, as well as other sometimes intangibles, such as violence, fear and traffic volumes that preclude children accessing outdoor spaces. The focus is on the inner city, but we recognise that some of the restrictions and challenges we identify can apply in wider spatial contexts of deprivation. In addition to recognising the challenges, we also identify opportunities and ways that even the most challenging environments can be nature enhanced.

The presence of nature in children's everyday environment is vital to their well-being. For children experiencing deprivation, the challenge is to provide nature opportunities as part of their daily lives, around their homes, streets and neighbourhoods. In densely built up areas, it may not be possible to easily provide opportunities at the greenest end of the continuum (Figure 1.8), but it can be done perhaps in a few smaller spaces. Examples used in this chapter provide hope and evidence of what can be done. All children, regardless of location and socio-economic condition, have a right to enjoy the benefits that come from having nature in their lives; the challenge is how to achieve it.

Deprivation: lack of nature versus lack of access to nature

Children growing up in high-density inner cities experience a number of advantages associated with access to city facilities but can also suffer a multitude of disadvantages related to difficulties in accessing outdoor green spaces. Inner-city environments are largely dominated by functional spaces and economic imperatives, commercial and retail spaces, high-density residential areas and spaces associated with high-density vehicle and people movement. Space is at a premium, and natural space usually loses out in the economic stakes. Consequently, city children growing up in the inner city can experience a lack of green access. Green space distribution in cities is uneven, with less green space generally associated with lower-income areas (Australia, Astell-Burt et al., 2014) and areas with higher proportions of some ethnic groups (African Americans, Hispanics and Asians; Dai, 2011). Higher proportions of public housing are often places where green space, especially private or domestic green space, is lacking or of poor quality.

The green-space-socio-economic status relationship for neighbourhoods is not a simple linear relationship, as some inner cities can be green-space-rich when there are large parks or where de-industrialisation has led to an increase in green space provision and spontaneous natural recovery. Parks can help combat deprivation. In Antwerp, the regeneration of old railway land at Parc Spoor Noord created a new green space that has helped reduce poverty and the high level of deprivation in the neighbourhood (European Commission, 2012). Older public parks tend to be located in areas close to city centres as in Hyde Park, London; Bosques de Palermo, Buenos Aires, Argentina; Luxembourg Garden, Paris; and Parque O'Higgins, Santiago, Chile. All of these parks contain a mix of formal landscaped areas and less formal habitats of greater ecological value. In the inner city, Central Park, New York, abuts Harlem, a traditionally low-income neighbourhood, and in the UK, parks often dating back to Victorian times are centrally located and thus close to traditionally low-income working-class neighbourhoods. Ironically, it can be the newer suburbs on the city fringes that are the most deficient in public green space. Public spaces are vitally important for children in the high-density neighbourhoods, especially where private green space around the home or apartments can be scarce or absent.

The social dimension

A sense of place is essential for children in accessing their nearby neighbourhoods, for playing outside and in encouraging use of public green space. A strong sense of place can also act as a motivator for use of existing spaces and for green space regeneration programmes. External assessments of sense of place can be masked by messages of deprivation and negativity. These negative messages do not necessarily match the views of the children and young people who live there, who may value their neighbourhood living environment despite its problems. In their study of children living in inner-city council estates in the UK, Reay and Lucey (2000) found that yes, the children were aware of the problems their neighbourhood presented but, more important, that

> Unlike the uniformly depressing, pathologising sentiments about 'sink' council estates and their tenants conveyed in both government documentation and media reports on social exclusion, the children we interviewed communicated a far greater sense of both hope and survival.
>
> (p. 424)

This sense of valuing what outsiders may see as a deprived area was also evident in some of our earlier work on social connections. When we interviewed children living in what media often portray as the most problematic state housing area in our hometown of Dunedin, the children spoke positively of its sense of community, their freedom, the green spaces within the housing estate and the open spaces around it. They had stronger feelings of attachment and greater personal freedom than most children from the more 'desirable' suburbs (Freeman, 2010). This sense of commitment to place evident in challenging environments was also present in interviews we undertook with children in the Pacific (Freeman et al., 2015). Children in Suva, Fiji, highly valued Suva Point Beach (Figure 7.6a) even though it was littered with rubbish and was a muddy estuary, rather than being one of Fiji's famed white sandy beaches for tourists (Figure 7.6b). Nevertheless, Suva Point was their beach and close to their homes. For children, a sense of belonging is an essential prerequisite if spaces are to be used and if nature regeneration and greening programmes are to be successful. The benefits of access to green spaces have been clearly identified and particular benefits noted for children in deprived or densely built environments. Physical activity is enhanced by the presence of green space. Larger areas with more trees were found to be positively correlated with the health-related quality of life in Houston, Texas, for Hispanic children. As a population, Hispanic children experience higher levels of obesity, inactivity and deprivation than other ethnic groups (Kim et al., 2016). Another inner-city study of a large public housing development in Chicago, US, found that for girls, green space around the home could help them lead more effective, self-disciplined lives. For boys, the more distant green spaces were equally important (Faber Taylor et al., 2002, p. 49).

Green space access is associated with physical and mental health benefits for children, and its development needs to go hand in hand with addressing social concerns. An Australian study of who uses urban parks concluded, "[M]easures to increase people's connection to nature could be more important than measures to increase urban green space availability if we want to encourage park visitation' (Lin et al., 2014, p. e87422).[1] Brisbane, where the study was conducted, is generally well endowed with green space, yet 40% of those in the study did not use the parks. They found that adults expressing a higher nature connection are more inclined to use parks and their own gardens. They believe this finding is important for children too, as direct nature experiences are precursors for developing the adult nature connections exhibited by the more frequent park users in their study.

There are a number of factors that influence use of parks and green spaces. Fear and safety is a recurring theme in limiting children's activity and play and thus is a primary factor needing to be addressed if green space use is to be supported where safety concerns preclude use. Studies from across different cities indicate the constraints that can be placed on children in this regard, as illustrated through the following examples.

A study of 20 US cities found maternal fears and depression in particular led mothers to inhibit their children's outdoor play. Some 34% of the interviewees were fearful of their children playing outside (Kimbro and Schacter, 2011). In a Chicago study of low-income African American women, mothers reported similar fears around letting their children play outdoors, citing the presence of stray dogs, drug pushing (often associated with parks), fear of kidnapping and other antisocial behaviours as concerns. These concerns can be seen as reasonable given that in 2010 some 21 murders and 435 aggravated batteries were reported to police for their neighbourhood (Jarret et al., 2013). A related study of physical activity for children and youth in a disadvantaged, walkable neighbourhood (10/10 on the walkability index) with good open space provision reported similar restrictions on children's outdoor play related to fear of violence (Holt et al., 2009). Again, for this neighbourhood, the fears are not unfounded. In 2005, the year of the study, there were 199 violent crimes and 741 crimes against property for the neighbourhood.

In a very different environment, Johannesburg, South Africa, fear again dominates attitudes to green space for children from very different socio-economic backgrounds (Kruger and Chawla, 2005). Children reported feeling insecure in Joubert Park, as it was associated with

> hobos, street children, strangers from local flats who threatened and hurt them . . . [children say] . . . the park is not clean. Kids are harassed when playing there. Also, some children get raped there. But still we would like to play in the park.
>
> (p. 89)

Other concerns children in the study had around open spaces were people drinking, fights, litter, drug dealers, people smoking dope, people having sex in the bushes, harassment of girls by men and general crime (p. 92). Many of the children interviewed live in high-density housing, apartments, homes with multiple adults and buildings shared by multiple families. Domestic gardens are scarce as is green space around the home; yet, available open spaces vital for play are unused because of parents' and children's safety fears. The provision of safe green spaces close to the children's homes should be a priority, together with ensuring children's safety in parks and green spaces. For children, social factors thus need to be considered in any greening provision, especially for children living in environmentally and socially challenging environments. Next, we look at more positive responses to the greening challenge for children living in challenging environments. For many of these initiatives, creating places where children feel safe in using the space has been fundamental in changing spaces from neglected to used.

Redressing the nature imbalance in challenging environments

Nature can be accommodated in numerous ways, even in what at first sight seems to be an alienating environment with little green space or when the green space present has little natural interest (e.g., mown grass). Restoration of nature can occur on a number of different types of sites:

- Brownfield sites, which are sites that have in the past been developed and may be in the process of re-development
- Unused sites, i.e., land with a non-green primary purpose but that can also support natural features, such as canals, railway sides, footpaths
- Ecologically challenged sites, areas characterised by poorly developed ecosystems, such as playing fields, extensive lawns and bare ground
- Hard surfaces, sites dominated by asphalt or other hard or impermeable surfaces
- Child-oriented spaces, such as schools or community playgrounds

Even in densely built up areas, there are invariably neglected pockets of land available that can be transformed into pocket parks or spaces that can be planted and greened (Figure 10.1). Some of these sites, when redeveloped, can

Figure 10.1 A wildflower meadow and path created on a piece of 'leftover' land, Sheffield, UK.

be impressive sites that are destinations for children, or they can be sites that permit informal play and exploration, perhaps en route to school, or a patch of land behind their apartments. Here, we explore a number of initiatives that are directed at high-density living, that restore nature to deprived housing areas, provide rural surrogate experiences through city farms and restore nature to schoolyards. These examples are but a small sample of many such initiatives but do show what can be achieved in especially challenging and neglected environments.[2]

High-density housing in Europe and Asia

The social inhibitors of outside activity need to be addressed together with some of the physical inhibitors. So far, the reference to safety has been largely associated with societal safety concerns, but another major concern is traffic and physically dangerous environments. There have been strong calls from researchers, practitioners and organisations, such as 'Reclaim the Street' and 'Living Streets', to reduce the dominance of motor traffic, especially in residential neighbourhoods. The origin of these liveable street ideas lies in the Dutch concept of 'woonerf' or living street. Woonerf support shared spaces with traffic-calming strategies and low speed limits that, together with landscaping, act to prioritise streets as a pedestrian space. For children, especially very young children, street space adjacent to the home is the most accessible outdoor space. If the street space is considered dangerous, it can preclude children from journeying out and going further. The 'woonerf' concept has been taken up internationally and in the UK is characterised by the home zone initiative, which aims to reduce traffic through shared space design. An early proponent in the UK was The Methleys (Leeds), a high-density area of terraced streets. In 1996, one street was turfed over for a weekend, and local families came out to play and socialise on the 'grassy' street. The initiative gained significant media coverage, which enhanced awareness of alternative streetscapes. In the UK, home zone initiatives took off in the ensuing decade, meeting with varied success (Gill, 2007). For the most part, home zones have failed to include any natural landscaping element, focusing predominantly on hard rather than soft landscaping. This hard focus is in contrast to the newer Dutch woonerfs, which do tend to include natural landscaping (Figure 10.2a and 10.2b). A more positive UK example is Bonnington Square,[3] which is the product of community action. Bonnington shows how planting can be used to enhance pedestrian safety and small spaces: the central square can be used to create interesting, safe natural landscapes and play spaces. Many of the plants are New Zealand rather than UK species, reflecting the presence of New Zealanders in the square, who, like many refugees and immigrants, bring to their new place familiar plants and species (Figure 10.3a and 10.3b). The original community who revived the square talk with pride of creating a green and adventurous environment for their children in the very heart of London.

High-rise developments present a greening challenge at the 'highest' and most confronting level.

Figure 10.2a Northmoor, Manchester, UK. Home Zone with minimal natural landscaping and retaining car dominance.

Figure 10.2b Javaeiland, Amsterdam, the Netherlands. Dutch woonerf with a mixture of formal and more natural landscaping in a car-free environment.

Figure 10.3a Bonnington Square, roadside planting.

Figure 10.3b The redesigned central park in Bonnington Square, showing natural planting in a safe environment.

In Europe, high-rise has largely been associated with deprivation, poverty and lack of green spaces. While high-rise developments can be alienating for children, in themselves, they need not always mean a nature-deprived life. In Eastern Europe, communism was associated with the building of high-density, uniform housing. The communist housing blocks (Figure 10.4a), now refurbished, were

Figure 10.4a Communist-era tower blocks, Bratislava, Slovakia, now refurbished.

Figure 10.4b Walking in the forest, near Bratislava, Slovakia, three generations. The youngest is inside his father's coat; only his hat is visible.

the childhood home for Jozef. Although the housing and environs were basic in design and possessing little natural variety, trips to the local forest with his grandfather to collect mushrooms were a memorable part of his childhood. On his return to Bratislava in 2016, Jozef took his own son, together with his son's grandfather, back to the forest for a walk, continuing the family's association with this natural landscape (Figure 10.4b). This example indicates that some families can contest the restrictions imposed by built environments. High- and medium-rise blocks, such as those in Bratislava, continue to be home for many European children, especially those dependent on public sector housing (Lizon, 1996). Unfortunately, little attention continues to be paid to the landscaping of the space around the blocks, which are dominated by uninspiring, ecologically limited landscaping, comprising isolated trees and grass.

Land shortages mean high-density housing in block forms[4] are common too in the Netherlands, but here, there is a much greater and increasing emphasis on creating nature-conscious living environments. In the New Amsterdam suburb of IJburg, children can easily access natural habitats along the water's edge (Figure 10.5a). Elsewhere in Amsterdam, older higher-rise housing blocks are gradually being redesigned to encourage residents' connection to outdoor environments (Figure 10.5b). Vegetable growing on land around housing blocks is occurring, not just in the

Figure 10.5a New housing blocks, IJBurg, Amsterdam, encourage children's access to waterfront habitats.

Figure 10.5b Land around older housing blocks is being developed into community gardens.

Netherlands but also in the UK. The Birmingham inner-city high-rise block in Figure 10.6a and Figure 10.6b has diversified, with vegetable boxes at the base of the block and self-sustaining wildflower species on common ground around the blocks. In addition, the canals adjacent to these tower blocks over time are being rehabilitated and now support increasing numbers of plant, insect and bird species. If children continue to be housed in high-rise blocks, initiatives such as these are a step forward, albeit only a slight one, in redressing the nature alienation inevitably associated with high-rise living.

While high-rise has been unpopular in western countries and until now was rare in countries such as Australia and New Zealand, it is becoming more common in major cities, such as Auckland and Melbourne. In Melbourne, between 1981 and 2001, multi-unit dwellings increased by 53%, and it was projected that in 2012, some 10,000 children aged under 15 would be living in the city of Melbourne,[5] most in high-rise homes (Whitzman and Mizrachi, 2012). In building these high-rise homes, the developers' assumption is that they will be occupied by DINKS (dual-income, no kids) or empty nesters (post-family stage), and they thus ignore provision for the needs of the families that increasingly are living there. Play space and open space adjacent to and overlooked by these units is conspicuously absent, making these potentially very alienating environments for young children. The

Figure 10.6a Vegetable growing adjacent to a high-rise housing block, Ladywood, Birmingham, UK.

Figure 10.6b Wildflower planting around the same tower block.

issue in Melbourne is not high-rise in itself, but poor design, a major flaw given the inevitability of their increasing prevalence in Australia and their growing use by families. Similarly alienating environments are being constructed in inner-city Wellington in New Zealand, with equal disregard for the play and open-space needs of the children who are also increasingly living there. The *Natural Neighbourhoods for City Children* study found the children living in this housing location of high-density apartments to be the least nature-connected of all the children in the study (Figure 10.7).

In countries such as Singapore and Hong Kong, high-rise is the dominant housing form. Providing for nature is a major challenge when population densities are 5,385 per km^2 (Hong Kong), up to 26,950 on Hong Kong Island (4 to 600,000 in some blocks) and 9,500 per km^2 in Singapore (Yeh and Yuen, 2011). Given the context of limited land and increasing populations, the challenge of providing nature connection opportunities is immense, with tall buildings being the most "unecological of all building types" (Yuen, 2011a, p. 181). The high-rise policy which 'frees' land for open space can be seen as a rational response to meeting residential needs, whilst maximising open-space provision (Yeh, 2011). Unlike in the UK, France, and some other European countries, in Singapore and

Figure 10.7 The bold colours and lone cabbage tree do little to offset the alienating environment children living in these apartments will experience, Wellington, New Zealand.

Hong Kong, there is general acceptance amongst the population of high-rise living, with residents preferring to live above the fifth floor where it is seen as less noisy with better air and views (Yeh and Yuen, 2011). High-rise living can never provide children with a nature-centred living environment, but given the propensity for high-rise living in Hong Kong and Singapore, the question is how can nature be maximised in such environments? Despite the predominance of the built, large areas of Singapore and Hong Kong are green. Hong Kong has about 73,300 ha of woodland, shrubland and grassland accounting for about 66% of the territory's total area, but its distribution is uneven, as it is found mainly at higher altitudes and away from the urban core. Similarly, significant land areas in Singapore are open space at around 24% (Yuen, 2011b). Singapore is prioritising green space provision such that by 2020, it plans to provide parks to within 400 m of all homes and 360 km of park connector spaces and have in place a substantive open-space provision and naturalisation programme (Urban Redevelopment Authority, 2012).

For Singapore children, the most important spaces are those contiguous to their homes. Those least able to access open spaces are the very young. For some older children, restricted access to natural space in the home environment can be offset to a degree by greater freedom of movement, enabling access to places further away. For young children, internal courtyard spaces and communal spaces, such as decks and roof spaces, are vital. However, the main provision in high-rise blocks centres on play equipment provision and formal, often hard-landscaped areas rather than on more natural plantings. Even these types of green space provision can have limitations, as young children generally need to be accompanied in order to use these open spaces. Time spent is necessarily dictated by adult availability and free time, with each visit to the play area estimated to last only 20 to 30 minutes (Yuen, 2011b). Pyle's (1998) concept of the 'extinction of experience' is particularly relevant for children in vertical living spaces. If parents had limited experience of outdoors, expectations will be limited for their own children. On the positive side, greening and restoration of nature in open spaces in Singapore, through internal courtyards, roof gardens and enhanced open space landscaping, are positive steps in the right direction. However, high-rise living can never be natural living when the nature of high-rise means that there will be no or minimal wild species present in the child's living environment, and any nature that is there will be highly managed.

Restoring nature to existing green spaces

Sheffield has been much studied as a city in terms of urban nature. It has been the focus of groundbreaking research, particularly on garden wildlife (Gaston et al., 2005). This early research was fundamental in drawing attention to the wildlife potential of urban areas. Sheffield though has its challenges in that distribution of green space is highly varied and uneven. A study by Barbosa et al. (2007) found,

> Only 36.5% of urban households in Sheffield meet English Nature's recommendation that each person should have green space no further than 300 m or

> 5 min walk from their home . . . 95.6% of the households meet the European Environment Agency recommendation of access to green space within 15 min walking distance, which is roughly equivalent to 900 m (p. 190) . . . Further, the distance to the nearest municipal park from some households in Sheffield is more than 20 times the EN recommendation.
>
> (p. 192)

While the overall scenario seems at first glance to be positive for a city characterised by high levels of relative deprivation and known as an industrial city, when looked at from the perspective of children's access, the outlook is less promising. Children's mobility has decreased markedly over the last three generations (Woolley and Griffin, 2015), such that few children have freedom beyond their immediate neighbourhood and some beyond the immediate vicinity of their home. To be accessible and used, the most important green spaces are private gardens, but these are scarce in the inner city, and public green spaces need to be as close as possible to the home and be enticing. Many children are not allowed to have independent access to spaces 900 m or further, and for some, even 300 m (English Nature's recommended access) would be too far, especially given prevalent safety concerns. In this regard then, the Sheffield Living with Nature programme is a vital element in supporting Sheffield children's nature connections (Woolley et al., 2014).

Successful nature restoration programmes can be important catalysts for enhanced community development and ideally work on multi-layered projects providing spaces that work for different ages, provide different habitat types and integrate different types of spaces. The Sheffield Living with Nature programme links redevelopment of green spaces, focusing on natural play with community projects. Most of the projects are in low-income neighbourhoods that are often state[6] housing areas. The Sheffield site redevelopments have included wildflower meadows, urban art with a nature theme, natural materials for play (such as boulders), incorporation of changes of level/slopes into the landscape and natural wood seating. On other sites, vegetable planters, nature walks, bird box construction, wildlife gardens, bulb planting and nature walks have been built (Figure 10.8a). For children, the priority was to create stimulating and interesting play experiences through using natural elements – boulders, logs, trees, mounds, ditches and landforms generally – locating equipment amongst natural elements whilst improving biodiversity on sites through creation of a range of habitat types (Figure 10.8b). Community and particularly children's involvement was an essential component of the process, as it creates a sense of responsibility and ownership of the sites and thus enhances their potential use. Commenting on the project, Karen Ainsley of Netherthorpe Primary School said,

> Our children have been involved in many activities run by Living with Nature which has enabled many of the children to plant and care for a living area which they have never experienced before. By having such ongoing involvement in the area, the children have seen it as part of their community and something they have an investment in and responsibility for. Many of our

Figure 10.8 (a) A swing on one of the Living with Nature sites and (b) Part of a revitalised play area that includes natural elements and natural landforms, such as slopes.

> parents are new to the area and indeed to the UK, having a focal point for the children to play means that the new parents have an opportunity also to engage with their new local community.
>
> (Woolley et al., 2014, p. 38)

Netherthorpe, which is one of the communities benefitting from the programme, is a multi-racial community of around 3,000 people (84.2% of schoolchildren come from black minority ethnic groups). Located in inner-city Sheffield, it is rated as a 'Most Deprived Area' with 30% of households claiming income support and 75% of residents living in flats, maisonettes or apartments, mostly rented from the local authority. In this context, the greening programmes have been immensely visionary and a valuable contributor to community development (Sheffield City Council, 2013).

Animals and vegetables: from city farms to vegetable plots

City farms, while not usually classed as nature sites, are, nonetheless, sites where children can get close to animals and provide, in some respects, a surrogate rural-country experience. Farms provide opportunities to meet animals in semi-natural settings, such as sheep in paddocks or ducks in ponds. They are increasingly including an ecological component with an environmental education focus. There are many examples of city farms, but Ouseburn City Farm in Newcastle portrays many city farm features. It has served its inner-city community for forty years, contains a productive garden and has an orchard. It was started by local people who wanted children living in the city to have the chance of working with farm animals. It has had its challenges, closing in 2002 when it was found to have contaminated soil, reopening in 2009, and recently again threatened with closure because of lack of funds. In addition to farm-based activities, there are a number of activities focusing on wild nature, including a wildflower hunt, spotting butterflies and dragonflies, pond dipping, hunting for mini-beasts and building a bug hotel. Entry to the site is free for visitors and, importantly, is easily accessible to children living in the Byker area. The Byker is known for its high levels of social deprivation, although redevelopment is aiming at redressing its associated social decline (Your Homes Newcastle, n.d.). For children and families on the Byker estate, the city farm presents a valuable opportunity to engage directly and locally with nature (Figure 10.9a and Figure 10.9b).

While city farms have usually focused on animals, many are diversifying with a growing focus on urban ecology and habitats and on providing city children with opportunities to practice 'farming' through growing plants. In Amsterdam, Jeugdland Maakland, a multifunctional site on the edge of Amsterdam, provides a full range of experiences for city children, including hut building, water activities, farm animals, indoor activities, natural picnic and play areas with a pond and plots where children can grow vegetables (Figure 10.10a and Figure 10.10b). Closer to the city centre, a city park has dedicated areas where children have their own

Figure 10.9a The vegetable section at Ouseburn City Farm.

Figure 10.9b Two local boys watching cattle at Ouseburn City Farm.

Figure 10.10a Jeugdland Maakland, Amsterdam, showing the children's vegetable plots adjacent to the playground.

Figure 10.10b Families at Jeugdland Maakland.

Figure 10.11 Children's named vegetable plots in an Amsterdam park.

plots for vegetable growing, which not only enables children to directly engage with the natural life cycles of plants but also gives them a stake in and sense of ownership in the park. Figure 10.11 shows children's named plots. Vegetable growing has also been one of a number of initiatives undertaken in school greening programmes.

Schoolyards

All children attend school, and all schools have some open space and the capacity to facilitate children's nature experiences through the school curriculum and through school trips. For some of the children in our *Natural Neighbourhoods for City Children* study, the school trip was their only camping trip or trip to the beach. This was the case even though two of the neighbourhoods of some study participants were located within walking distance of the beach. For some children, the school day trip to the beach was their first experience of that beach. Given the limitations placed on some children's outdoor play and independent mobility, schools can be an important conduit for nature experiences. This is particularly so for children whose lives lack nature enrichment experiences, such as visits to forests, or those who lack access to domestic gardens or accessible public green space. The UK organisation Learning through Landscapes estimates that 'Most children and young people will spend around 1500 hours of their life in their school playground. For many, this will be more than in any other outdoor play setting' (Learning through Landscapes, n.d.). It is essential, therefore, that these hours be spent in an educational and enriched natural setting. Ironically, the schoolyard can still be many children's least natural area, given the predominance of asphalt and hard surfaces.

Where school grounds contain more natural biophilic elements, they can encourage more active play and development of gross motor skills, as well as supporting affective, cognitive and socialisation skills (including conflict reduction) and development (Gilchrist et al., 2017; Malone and Tranter, 2003). A number of studies have backed these findings. A US study of children aged 4 to 8 years found "children's play activities were more sustained, constructive and cooperative when given the affordances of a natural playscape" (Kuh et al., 2013, p. 70). Other findings have been more mixed. One US study found a positive relationship between reading performance and tree cover (Hodson and Sander, 2017). There is a well-established preference for children electing to spend time in natural parts of the schoolyard where these are available (Lucas and Dyment, 2010). Children therefore prefer natural schoolyards, but schoolyard design, especially in the inner city, is still often dominated by concerns related to functionality (e.g., surveillance, maintenance, weather-proof, robustness, cost, sports use), leading to the predominance of large asphalt areas or monocultural grass fields (see number 1). Four types of school playground can be identified:

1 Traditional – dominated by asphalt and sports fields, focusing on space being used for physical exercise
2 Designer – aesthetics introduced but in ways that predetermine play activities through fixed equipment
3 Adventure – includes a range of natural spaces and materials, limited predesigned elements
4 Creative/comprehensive – synthesis of types 1–3, exercise areas, sports fields, equipment and natural settings.

(Malone and Tranter, 2003)

Recognition of and encouragement of more natural schoolyards has been actively supported in recent years by a number of organisations, such as Green Schoolyards America, who have produced a number of activity guides for schoolyards (greenschoolyards.org). ReForest London in Canada have planted trees in over 30 schoolyards, involving at least 3,000 schoolchildren. In London (UK), the London Children's Flowers Society has provided planter boxes to schools across the city (London Children's Flower Society). One of the longer established organisations, UK Learning through Landscapes have also produced a *Good School Playground Guide*, which promotes natural materials and natural play. These initiatives are but a sample of many school ground initiatives occurring globally. In New Zealand, where space is not at such a premium, even inner-city schools tend to be well endowed with natural spaces. School grounds tend to include natural elements, such as trees, slopes and shrubs as well as asphalt and sports fields and increasingly vegetable-growing and natural habitats. These developments have been supported by the strong Enviroschools movement. Enviroschools works with 1,000 schools and some 260,000 children and young people to provide a more environmentally appropriate and sustainable experience, supporting natural school grounds and environmental experiences (Enviroschools, 2017).

Schools are further diversifying their grounds with the development of wildlife gardens. There are organisations which promote and assist with the development of such gardens. One such guide produced for an inner London (UK) borough gives guidance on creating meadow habitats, bog gardens/marshes, log piles, bird tables and feeders, water baths, bird boxes, stone/brick walls, herb gardens, butterfly gardens, bat boxes, wormeries and compost heaps (The Ecology Service, n.d., Figure 10.12a and 10.12b). Schools can be an important catalyst for nature experiences and an especially vital counterpoint for children whose lives are otherwise lacking in nature experiences. While the challenges in restoring nature to the school ground can be considerable in some schools, especially in the inner city or schools with limited outdoor space, all schools can introduce natural elements in both play and for attracting wildlife, even if only through using bird feeders and

Figure 10.12a A simple compost and recycling arrangement at a school in Auckland, New Zealand. On the top wire is a fantail attracted by the flies.

Figure 10.12b A school flower and vegetable garden, Devonport, Plymouth, UK.

planters, introducing trees and making better use of the green spaces close to the school (ReForest London).

Bringing nature into children's lives

In this chapter, the focus has been on spaces that can make a notable difference for children in high-density or deprived living environments. Though the focus was on enhancing nature provision in four specific spaces – high-density high-rise, low-rise high-density estates, city farms and schoolyards – these are but indicative of how spaces can be enhanced. The enhancement principle can be applied throughout the city. Biophilic design is described by Beatley as follows:

> A biophilic city is a green city, a city with abundant nature and natural systems that are visible and accessible to urbanites. It is certainly about physical conditions and urban design-parks, green features, urban wildlife, walkable environments . . . it is also about the spirit of a place, the emotional commitment and concern about nature.
>
> (Beatley, 2011, p. 17)

Biophilic design can and should be applied throughout the city and especially in children's living environments. In this chapter, we have focused on environments in which children spend considerable periods of their life, but these need to be part of a biophilic city, not isolated patches of biodiversity in an otherwise nature depauperate environment.

To end this chapter, it is instructive to refer to three examples that reveal how more natural biodiverse environments can be purposefully designed into otherwise inhospitable environments. These examples identify created places that act as magnets for children and their families. The first is the High Line, a disused rail line in New York's slaughterhouse district that has been transformed into an aerial greenway with its own ecological habitats of meadows, grassland and woodland, comprising a mix of native and exotic species (La Farge, 2012). It is much used by families for outings. Phase 3 of the development includes specifically designed play and adventure elements for children that are built around the High Line's ecological elements.

A second example but one that is both opportunity taken and opportunity lost is that of the Cheonggyecheon stream (Figure 10.13a), an 11-km stream running through central Seoul. The stream had been hidden in a drain under an elevated highway. Reinstated at a cost of $900 million, the project was controversial,

Figure 10.13a Restoration of the Cheonggyecheon stream, central Seoul. The stream acts as a magnet for families (the man lower left is pointing to the large fish that inhabit the stream). In central Seoul, the priority is amenity rather than ecological restoration.

Photo by Jinty Mctavish

Figure 10.13b Habitat restoration of the Rive Aire and canal Leeds, UK.

attracting much public criticism during construction. Since opening in 2005, the stream has been promoted as an example of successful rehabilitation, but is it? On the positive side, it is a magnet for commuters, shoppers and families, who can be seen using the walkways even into the evenings. The stepping-stones across the river are a particular favourite for children, as from them, they can see the stream's fish. It therefore has been hugely successful in terms of amenity value. Its ecological benefits are less straightforward. The restoration of the stream has

> Increased overall biodiversity by 639% between the pre-restoration work in 2003 and the end of 2008 with the number of plant species increasing from 62 to 308, fish species from 4 to 25, bird species from 6 to 36, aquatic invertebrate species from 5 to 53, insect species from 15 to 192, mammals from 2 to 4, and amphibians from 4 to 8.
>
> (Landscape Architecture Foundation, 2017)

Ecologically, it could be argued that despite these seemingly impressive figures, the stream restoration is a missed opportunity. The central and most used parts of the stream are the least ecological, and its maintenance is dependent on continual and invasive management including pumping 120,000 tonnes of water into the stream daily to keep it flowing.

A more ecologically appropriate example and one referred to in Chapter 5 is the River Aire in Leeds (Figure 10.13b). Running through the centre of the industrial city of Leeds, "by 1825 the river was nearly devoid of life and remained so for another 100 years" (Environment Agency, n.d.). Now the river and the adjoining canals have been rehabilitated. In the recent past, the river at times of low rainfall comprised 70% sewage effluent and just 30% natural water (Lazenby, 2011). Today, birdlife, including herons, salmon and even otters are returning to the river in the heart of the city.

Most cities contain 'forgotten space', where natural habitats can be encouraged (or where species self-seed and thrive relatively unnoticed on leftover pieces of land, such as canal sides, or derelict land. All three examples, the High Line, Cheonggyecheon stream, and the River Aire, were 'forgotten spaces' in their city centres. These overlooked spaces have immense potential for combating the nature deficit in high-density environments. These are but three examples of how biodiversity in overlooked spaces can be brought into cities and children's lives, in what are otherwise deeply challenging, biodiverse deficient environments.

Notes

1 Brisbane Australia is well endowed with green space so is not typical of more densely urbanised cities, but the message is, nonetheless, apposite.
2 See also examples such as Camley Street Natural Park in chapter 11.
3 Bonnington Square, https://vimeo.com/36595608.
4 Although the housing block form is common, the blocks tend to be well designed and are more architecturally interesting than those characteristic of communist Eastern Europe.
5 I.e., inner suburbs.
6 Government-funded housing for low-income residents.

References

Astell-Burt, T., Feng, X., Mavoa, S., Badland, H.M., and Giles-Corti, B.L. 2014. Do low-income neighbourhoods have the least green space? A cross-sectional study of Australia's most populous cities. *BMC Public Health* 14: 292.

Barbosa, O., Tratalos, J.A., Armsworth, P.R., Davies, R.G., Fuller, R.A., Johnson, P., and Gaston, K.J. 2007. Who benefits from access to green space? A case study from Sheffield, UK. *Landscape and Urban Planning* 83: 187–195.

Beatley, T. 2011. *Biophilic Cities: Integrating Nature into Urban Design and Planning*, Island Press, Washington.

Dai, D. 2011. Racial/ethnic and socioeconomic disparities in urban green space accessibility: Where to intervene? *Landscape and Urban Planning* 102(4): 234–244.

The Ecology Service. n.d. *The Ecology Service: School Wildlife Gardens: A Teacher's Guide*, The Ecology Centre: Royal Borough of Kensington and Chelsea, London. Accessed December 2016.

Environment agency. n.d. *River Aire Fact File*. Environment Agency, Leeds. www.environmentdata.org/archive/ealit:1270/OBJ/19001000.pdf

Enviroschools New Zealand Tā Tātou Kaupapa About Enviroschool. www.enviroschools.org.nz/about-enviroschools accessed January 2017.

European Commission. 2012. http://ec.europa.eu/regional_policy/en/projects/belgium/urban-park-revitalises-deprived-city-neighbourhood accessed January 2017.

Faber Taylor, A., Kuo, F.E., Sullivan, W.E. et al. 2002. Views of nature and self-discipline: Evidence from inner city children. *Journal of Environmental Psychology* 22: 49–63.

Freeman, C. 2010. Children's neighbourhoods, social centres to 'terra incognita'. *Children's Geographies* 8(2): 157–176.

Freeman, C., Lingam, G.I., and Burnett, G. 2015. Children's changing urban lives: A comparative New Zealand–Pacific perspective. *Journal of Urban Design* 20(4): 507–525.

Gaston, K., Warren. P.H., Thompson, K., and Smith, R.M. 2005. Urban domestic gardens (IV): The extent of the resource and its associated features. *Biodiversity & Conservation* 14(14): 3327–3349.

Gilchrist, M., Passy, R., Waite, S., and Cook, R. 2017. Exploring schools' use of natural spaces. In C. Freeman and P. Tranter (Eds.), *Risk Protection, Provision, and Policy, Geographies of Children and Young People*, vol. 12, Springer Major Reference Work.

Gill, T. 2007. *Can I Play Out? Lessons from London Play's Home Zones Project*, London Play, London.

Green School Yards America. www.greenschoolyards.org accessed January 2017.

Hodson, C.B., and Sander, H.A. 2017. Green urban landscapes and school-level academic performance. *Landscape and Urban Planning* 160: 16–27.

Holt, N.L., Cunninghamn, C.T., Sehn, Z.L., Spence, J.C., Newton, A.S., and Ball, G.D.C. 2009. Neighbourhood physical activity opportunities for inner-city children and youth. *Health & Place* 15: 1022–1028.

Jarret, R.L., Sensov, B.O., McPherson, E., and Williams, D.A. 2013. No child left inside: The built environment and caregiver strategies to promote child physical activity. *Journal of Leisure Research* 45(4): 485–513.

Kim, J.H., Lee, C., and Sohn, W. 2016. Urban natural environments, obesity, and health-related quality of life among Hispanic children living in inner-city neighbourhoods. *International Journal of Environmental Research and Public Health* 13: 121.

Kimbro, R.T., and Schacter, A. 2011. Neighbourhood poverty and maternal fears of children's outdoor play. *Family Relations* 60: 461–475.

Kruger, J., and Chawla, L. 2005. 'We know something someone doesn't know . . .' Children speak out on local conditions in Johannesburg. *Children, Youth and Environments* 15(2): 90–104.

Kuh, L.P., Ponte, I., and Chau, C. 2013. The impact of natural playscape installation on young children's play behaviors. *Child, Youth and Environments* 23(2): 49–77.

La Farge, A. 2012. *On the High Line*, Thames and Hudson, London.

Landscape Architecture Foundation. 2017. *Cheonggyecheon Stream Restoration*. https://landscapeperformance.org/case-study-briefs/cheonggyecheon-stream-restoration Project

Lazenby, P. 2011. Yorkshire's revived river Aire is a lesson in people power. *Guardian*, Monday 5 September.

Learning through Landscapes. n.d. *The Good School Playground Guide Developing School Playgrounds to Support the Curriculum and Nurture Happy, Healthy Children*, Learning through Landscapes, Winchester. www.ltl.org.uk/resources/results.php?id=712

Lin, B.B., Fuller, R.A., Bush, R., Gaston, K.J., and Shanahan, D.F. 2014. Opportunity or orientation? Who uses urban parks and why. *PLoS One* 9: 1.

Lizon, P. 1996. East Central Europe: The unhappy heritage of communist mass housing. *Journal of Architectural Education* 50(2): 104–114.

London Borough of Camden. 2006. *Supplementary Planning Document Sites of Nature Conservation Importance in Camden*, London Borough of Camden, London.

London Children's Flower Society. *Planting in the Playground.* www.londonchildrensflowersociety.org/?page_id=54 accessed January 2017.

Lucas, A.J., and Dyment, J.E. 2010. Where do children choose to play on the school ground? The influence of green design. *Education* 3–13, 38(2): 177–189.

Malone, K., and Tranter, P. 2003. Children's environmental learning and the use, design and management of school grounds. *Children, Youth and Environments* 13(2): 87–137.

Pyle, R.M. 1998. *The Thunder Tree: Lessons from an Urban Wildland*, Lyons Press.

Reay, D., and Lucey, H. 2000. 'I don't really like it here but I don't want to be anywhere else': Children and inner city council estates. *Antipode* 32: 4410–4428.

ReForest London. http://usfn.ca accessed January 2017.

Sheffield City Council. 2013. *Statistics from Sheffield Neighbourhood Health & Well-Being Profiles 2012 Netherthorpe Neighbourhood Sheffield City Council Sheffield Neighbourhood Health and Well-Being Profile*, Public Health Intelligence team, Sheffield City Council.

Urban Redevelopment Authority (URA) Singapore. 2012. *Designing Our City Planning for a Sustainable Singapore*, URA, Singapore.

Whitzman, C., and Mizrachi, D. 2012. Creating child-friendly high-rise environments: Beyond Wastelands and Glasshouses. *Urban Policy and Research* 30(3): 233–249.

Woolley, H., Bradshaw, K., Somerset-Ward, A., Fagg, S. et al. 2014. *Stories from Living with Nature*, University of Sheffield, Sheffield, UK.

Woolley, H., and Griffin, E. 2015. Decreasing experiences of home range, outdoor spaces, activities and companions: Changes across three generations in Sheffield in north England. *Children's Geographies* 13(6): 677–691.

Yeh, A.G.O. 2011. Tall building living in high density cities: A comparison of Hong Kong and Singapore. In B. Yuen and A.G.O. Yeh (Eds.), *High-Rise Living in Asian Cities*, Springer, pp. 9–23.

Yeh, A.G.O., and Yuen, B. 2011. Introduction: High-rise living in Asian cities. In B. Yuen and A.G.O. Yeh (Eds.), *High-Rise Living in Asian Cities*, Springer, pp. 1–8.

Your Homes Newcastle. n.d. www.futurecommunities.net/case-studies/byker-estate-newcastle-1967-present accessed January 2017.

Yuen, B. 2011a. In the march of high-rise. In B. Yuen and A.G.O. Yeh (Eds.), *High-Rise Living in Asian Cities*, Springer, pp. 179–186.

Yuen, B. 2011b. Garden and ecology in the sky. In B. Yuen and A.G.O. Yeh (Eds.), *High-Rise Living in Asian Cities*, Springer, pp. 65–79.

11 Working for and with children for more biodiverse cities

What do we need for a biodiverse, child-friendly city?

Biodiverse cities do not happen by chance. They are the consequence of concerted actions that act to sustain and enhance their habitats and the species they support. Positive actions rely on the vigilance and commitment of the children and families living in them together with the vigilance and commitment and political advocacy of the professionals, agencies and organisations whose remit it is to manage and develop the city's biodiversity. The city and its biodiversity are not static but constantly responding to the pressures of building development, population growth and change, urban decline and regeneration and ecological modification, as well as the vicissitudes of its seasons, climate and geography. For children, access to nature is dependent on the actions of those with close personal ties to the child – family, relatives, teachers and friends – but also those whose professional remit determines the shape and function of the city for its inhabitants, both human and wild. For a city to work well for all its inhabitants requires coordinated vision, commitment, action and the flexibility necessary to respond to a dynamic environment. In this chapter, we examine what it takes to make a biodiverse city for children, including the role of government, and the strategies and planning tools that make it possible. We consider the professions, agencies, organisations and people involved and what makes for a successful biodiversity initiative that will work for children. Finally, we focus on the role of children themselves, including initiatives in which children take a leading role and are involved as not just biodiversity users but as researchers and contributors to helping create biodiverse city environments.

Good practice is illustrated through a selected number of international exemplars that show how the city's biodiversity can be supported to provide places where children can interact with and experience nature in their cities. This includes not just spaces that are normally considered as relevant for children, such as parks and schools, but also critical life spaces, such as hospitals. The focus is on initiatives that are part of or reflect children's everyday and lived experiences with nature. However, for some children, as is the case for many of the children using the Cape Town Environmental Trust project (CTEET, 2017), the nature-deprived environments in which many of the children live mean that programmes external to their daily living environments are necessary for nature connection. In this chapter, we focus on two types of initiatives – firstly, those that demonstrate how effective planning and action can enhance living environments that encourage nature contact, and secondly, places that play a significant role in all or some children's lives, including schools, playgrounds, hospitals and parks,[1] where enhancing the biodiversity present can make a big difference in nature contact opportunities and in enhancing children's own well-being.

Planning for nature: the need for a coherent and coordinated strategy

National planning

Many countries now have national biodiversity strategies, and invariably they include statements that recognise that biodiverse areas provide benefits for the population, especially children. In this discussion, we profile international exemplars of projects from quite different countries. The most successful biodiverse cities and neighbourhoods have support at four levels: national governmental level, the city level, the neighbourhood level and the project level. To work effectively at ground/project level and to make a difference to children's lives, it is necessary to start at the top. There needs to be a national commitment to urban biodiversity and to seeing children as an integral part of the urban biodiversity relationship.

This national-level commitment can be a direct government plan for biodiversity, such as Singapore's National Biodiversity Plan. This links to the Greening Schools for Biodiversity initiative, which includes families engaging with nature (Singapore Government National Parks, 2014). Multilevel and multi-sectoral links are important in providing continuity across levels and for providing the backing for projects at the implementation level, in this case, the schools. The forms that these projects take differs, but in all cases, the projects build on wider societal commitments to creating and enhancing children's biodiverse lives. For these and for all projects, plans and strategies matter. If the guiding plans and strategies include policies and commitments to children and biodiversity, then it is much easier for local governments and organisations to leverage support and resources for biodiversity programmes. Further, they provide support for having the needs of children considered in urban planning and development and for potentially drawing resources towards supporting projects.

Commonly, responsibility for encouraging biodiversity awareness and development of child-focused biodiversity programmes is routed through a country's national organisation for biodiversity; in New Zealand's case, this is the Department of Conservation. In recent years, the Department of Conservation has started to pay much more attention to working in the urban environment and to working with and on behalf of children living there. The *New Zealand Biodiversity Action Plan 2016–2020* proposes that by "2020, New Zealand will have involved 230,000 children in programmes connecting them with New Zealand's natural environment and inspiring them to take action for nature in their backyards and communities" (Department of Conservation, 2016, p. 44). This is a massive shift for an organisation whose responsibility it is to manage and sustain species, habitats and ecosystems, many of them gravely threatened. Working in highly modified landscapes, such as cities, and with children has not previously been part of its primary mandate. Given the already significant conservation challenges, as well as resourcing challenges in the conservation estate, including the national parks, the growing focus on children and urban centres is encouraging. The Department of Conservation has been increasing its focus on children in the context of environmental education and engagement in biodiversity initiatives or programmes. In this respect, it has much in common with Singapore's environmental education-oriented national plan.

City planning

Biodiversity strategies and programmes vary in their recognition of the urban, and it is at the city level rather than the national level that children's urban nature connections are more commonly recognised and provided for. The city of Freiburg, Germany (approx. 220,000 inhabitants), is one example of how better planning can help create greener living environments for children. Freiburg's plan builds on a 40-year commitment to more sustainable living and prioritises biodiversity in its planning. This prioritisation has resulted in the cityscape currently comprising just 32% built, 42% forest and 27% other green spaces. As a city, Freiburg has 5,000 ha of forestry and 600 ha of parks (City of Freiburg, 2016). To encourage biodiversity, it has introduced a number of ecologically supportive practices that enable more natural landscapes to thrive. One example is a reduction in lawn mowing (public spaces) from 12 to two times a year, which encourages meadow growth. The city contains 160 parks, but whereas many cities are well endowed with parks, Freiburg takes parks, and in particular playgrounds, to a new and biodiverse level by eschewing the standard pre-formed plastic playgrounds in favour of natural play features built with natural materials and located in green settings. Children and local residents are consulted on new developments and park and playground regeneration projects, and their input helps shape the final design (City of Freiburg, 2016, p. 17). This focus on play carries through to the city centre where the 'Bächle' (streams) were brought back to the surface from underground drains and now run in small open channels, providing play features for children on hot days (International Making Cities Liveable, 2017).

At the neighbourhood level, one of its best-known redevelopment projects is the Vauban neighbourhood. It was created intentionally as an environmentally sustainable community. Pedestrians and cycling are prioritised such that 70% of the residents do not own a car. Therefore, not only is land freed up that can be greened, but children's access to the outdoors is safer and enhanced. Vauban is the greenest area in the already green City of Freiburg. In designing the housing, attention was paid to retaining natural features such that 'Green spaces between the rows of houses guarantee good climatic conditions and provide play areas for children' (City of Freiburg, 2016). Freiburg and Vauban in particular show a new way for city living that integrates biodiversity into better living environments. This development owes its success to forward thinking, commitment to an integrated vision for housing, transport, playgrounds, energy, waste and all the elements that go towards creating enhanced living environments. Another factor in its favour was that Freiburg 'benefits from an unusually high concentration of specialist professionals working in sustainability, including the ICLEI (Local Governments for Sustainability), the International Solar Energy Society (ICES) and the City Mayors Foundation' (Thorp, 2014). Good planning, a visionary strategy that was flexible and living documents prepared the ground for recurrent innovative environmental practices.

Initiatives such as those of Freiburg, especially where the initiative is wide-ranging in scope and location, work to create more ecological neighbourhoods for children. We identify just some of the professions that might be involved in city-level biodiversity initiatives. For each city and for each initiative, a different set of respondents and a different set of participants may be involved at different stages. The professions, stakeholders and tasks indicate just some of those who may be involved in any project development and the type of contributions they can make (Table 11.1). Critical to all of the process though are children, who should be central participants across the actions of all the major groups.

Table 11.1 Participation in biodiversity initiatives, professionals and communities

Major groups	*Professions and stakeholders**	*Task – contribution*
Government, related agencies and professions		
Children	City Council	City and local-level advocacy, advice and support, decision-making
	Politicians	Decision-making, community action, resources
	Planning	Site location, development approvals, policy, plans, negotiation and coordination
	Housing	Housing management and provision
	Traffic engineering	Roading, access, cycle ways
	Engineers	Infrastructure, construction, safety
	Transport	Transport management and provision
	Parks, recreation, amenity	Green space management and provision
	Environmental agencies	Expertise and resources

Major groups	*Professions and stakeholders**	*Task – contribution*
Development, resource providers and professions		
Children	Businesses	Resources, support, advertising
	Builders	Construction
	Architecture	Design and development
	Lawyers	Land ownership, development, rules, regulations and plans
	Designers	Landscape, play, green space provision and design
	Finance	Resources to support development
	Project development	Project management
Conservation, landscape and natural resources		
Children	Conservation	Species, habitat protection and management
	Environmental education	Education provision
	Landscape architecture and ecology	Green space design and development, plantings, public spaces
	Horticulture	Plants
	Water – hydro	Streams, rivers, ponds
Community		
Children	Schools	Community support and education
	Parents	Community/family support
	Relatives	Community/family support
	Friends	Community/family support
	Neighbours	Community support
	Local organisations	Coordination, practical support, motivation, connections
	Community worker	Support, coordination of services

*The allocation of professions and stakeholders to different major groups can vary according to the initiative, e.g., environmental education can be located within council, within conservation organisations or within communities or schools.

Biodiversity initiatives

Creating successful projects

What makes for a successful biodiversity initiative, and how do you know that the project is going to work for children? Clearly, as Table 11.1 indicates, there are a wide range of organisations and tasks to be considered and supported in developing and managing any project. The Department of Conservation (Wilson, 2011), New Zealand, in its report *Effective Approaches to Connect Children with Nature*, identifies a number of best practice principles for connecting children with nature. The Department of Conservation's interest in environmental education and children and conservation has tended to be directed at activities in the conservation estate and experiences associated with New Zealand's endangered indigenous species, primarily birds. This 2011 report, however, takes its starting point from the following idea: "[S]tudies suggest that frequent, unstructured experiences in nature are the most common influences on the development of lifelong conservation values" (2011, p. 1). In this respect, it identifies the following criteria for success for biodiversity initiatives:

1 Make it relevant to everyday life.
2 Include families, communities and role models.
3 Provide opportunities for social connections.
4 Promote direct experiences.
5 Encourage free-choice learning.
6 Foster the role of active stakeholder.
7 Target real local issues.
8 Promote collective action.

These criteria provide an excellent basis against which to consider urban biodiversity projects directed at children. Particularly relevant is the first, with its reference to the everyday, and the third. Here, we feature a number of projects that address some if not all of these eight criteria for successful initiatives. The projects approach children in different ways. They include projects that attempt to redress the lack of nature contact opportunities associated with children's home environments, neighbourhood eco-regeneration and regional nature programmes, including programmes where children work as eco-partners and researchers. It is not possible to cover the full spectrum of projects and the diverse ways to include children, but the following provide an indication of some of the various ways nature contact can be encouraged through locally based projects. While we have tried to focus on projects that can be accessed within children's everyday environments, such contact, especially for children from nature-deprived environments, is not always possible.

Providing nature contact for children from environments deprived in biodiversity

For children who lack access to domestic green space in the home environment, their local neighbourhood, and lack the resources to independently access more distant biodiverse sites, formal and educationally based nature sites become especially important. Such sites are usually accessed through school trips. For many children, especially those in highly urbanised environments, these trips may be their only opportunity to experience quality nature environments. For children in the New Zealand study who attended low socio-economic schools, the school camp is often their only immersive nature experience. One school principal described how some children from her Wellington school never went to the beach outside of the school trip, despite Wellington's coastal location. At another school, in Auckland, the children walked the twenty minutes from their school to the beach for an environmental education class (Figure 11.1), and for some of them, it was their first trip to the beach. The presence of natural areas in a city or neighbourhood does not necessarily mean they will be accessed, and for children who do not access them, organised trips to local nature sites are vitally important. The next two examples, South Africa's Cape Town Environmental Education Trust and Camley Street Natural Park focus on providing nature opportunities for children who would otherwise miss out.

Cape Town Environmental Education Trust (CTEET)[2]

The location of Cape Town is important, as it is situated in the smallest and most threatened floral region on the planet.[3] Given the biodiversity value and fragility of this region, having children relate to this very special environment is crucial for its long-term survival. Within this region, CTEET provides opportunities for the children of Cape Town to experience the outdoors, in the hope that these experiences will foster an interest in nature that will be expressed as pro-environmental values and behaviours as adults. The vision of the CTEET is "supporting the preservation of Cape Town's unique and biodiverse natural heritage through education, training and conservation initiatives" (CTEET, 2017).

Activities include trips away from often socio-economically and environmentally challenging home environments. Many of these children are afraid of nature. Their home environments are often deprived of any meaningful green spaces. Among the 'black and coloured' population groups, there are many beliefs attached to animals that can accentuate fear of nature. For example, owls are believed to predict death and chameleons are said to come from the devil. Children of all age groups attend

Figure 11.1 School trips can be excellent ways to familiarise children with what is available to them locally.

CTEET's camps, from schools, church groups and clubs (Figure 11.2). Most live in residential areas with very little green space, and very few have gardens. While play spaces may exist in neighbourhoods, the main factor limiting access is safety, e.g., gang wars that can take place both night and day. Nearly 30,000 children have attended overnight camps to date. Encounters with nature are seen as encouraging pro-social behaviour and boosting children's self-esteem. The biggest challenge for CTEET is getting the children to the camps, i.e., money for the bus. CTEET gets funding from many sources, including international sponsors.

CTEET has links with and provides resources to more than 50 eco-schools, which is an international programme coordinated in South Africa by the Wildlife Society for Southern Africa (WSSA). The aim of the programme is to encourage environmental awareness and behavioural change within schools that in turn spreads to the neighbouring communities. The Conservation Leadership Programme is another initiative, focussed on children from eco-schools who are passionate about the natural environment and who would like to play a role in effecting change in their homes and communities. Children 11 to 12 years of age from these schools are nurtured by the programme for up to seven years while at school. During this time, they interact with CTEET every month and participate in community action, hiking and camping in wilderness areas and camping trips to conservation areas throughout South Africa. It is hoped that the experiences they have, together with career guidance and knowledge sharing, will in turn create conservation leaders of the future.

Figure 11.2 Children meeting a leopard toad at CTEET.

Source: http://cteet.co.za/, Used with permission from CTEET

Camley Street Natural Park[4]

Built on derelict land that once was coal yards, the nature park opened in 1986. Located 480 metres from Kings Cross station (London, UK) and the Eurostar international rail terminus, this 0.8 ha site is a wonderful example of how nature can be reintroduced into even the most seemingly inhospitable of environments. Camley Street Natural Park is especially poignant, as it was nearly lost in the redevelopment of Kings Cross in the 1990s. It now comprises a mosaic of woodland, wetlands and ponds (where children pond dip), bug homes, picnic areas, grassy banks, canal-side, meadows, scrub and the banks of Regent's Park Canal. Swans nest on the canal banks where they can be observed by visiting children (Figure 11.3). Managed by the London Wildlife Trust, entry is free, and the ecological regeneration on the site has led to it being designated a Site of Metropolitan Importance for Nature Conservation as well as a Local Nature Reserve (London Wildlife Trust, 2017). The site is extensively used, with approximately 20,000 visitors a year, including thousands of local schoolchildren, and it is accessed by local children, workers and families (Figure 11.4) (London Borough of Camden, 2006). The site is close to a number of large council estates (with public housing) characterised by low incomes and deprivation. Camden, the borough in which the nature park is located, has 36% of its children living in poverty (Child Poverty Action Group, 2017), with

Figure 11.3 Photographing a nesting swan from inside Camley Street Natural Park

Figure 11.4 Pond dipping at Camley Street Natural Park.

many children living in high-rise housing with minimal access to natural spaces around their homes. While there are some good open spaces in Camden, such as Primrose Hill, many low-income families would not ordinarily visit them. By becoming familiar with the nature park through school visits, local children and their families are much more likely to make return visits and see the natural park as part of their familiar local landscape.

Creating more biodiverse city spaces

For play, natural spaces are needed at different levels. These include city-level spaces, neighbourhood spaces and spaces that are part of daily life; these various spaces may be used more or less frequently, depending on location and resources. We provide examples of such spaces. At the highest level in a city are destination spaces. These will be designed to attract and provide for children across the city. Their use is usually dependent on the children having the resources to travel to the play space. As such, they can be important destinations for school visits by children who would otherwise be unlikely to access them. Even if the space is free on entry, as with the RioTinto Naturescape in Perth,

Australia, travel costs can be a barrier to access. In our New Zealand study, we found that living close to an educational opportunity, such as an ecosanctuary or zoo, did not mean the children accessed the space, likely because of a lack of resources. At the second level are neighbourhood spaces, usually a park, woodland or area of natural habitat. These spaces are especially important for children living in high-rise or housing that is compromised with regard to green space provision. Bishan-Ang Mo Kio Park[5] in Singapore has provided a green space for local residents since 1988, but its recent makeover transformed the park into a nature experience through a committed programme of nature restoration. Jeugdland Maakland, Amsterdam, provides a space where children can experience and restructure their own play space. On the third level are daily life spaces, such as hospitals; educational spaces, such as schools; streets and gardens. The following examples demonstrate how nature connection opportunities can be created at all these levels and the importance of enabling nature to be integrated into children's activity spaces.

These examples demonstrate the importance of thinking differently, of seeing, for example, a play space as being more than pre-formed plastic structures or a hospital as more than beds, waiting rooms and play equipment placed in a corner of the room. The success demonstrated in the following examples came about because the planners, designers, play specialists, health professionals, hydro engineers and other participating experts went beyond simple functional thinking. The hydro engineers at Bishan-Ang thought about how a river trapped in a canalised drain could be made safe in times of flooding by allowing it to meander and flood naturally, while simultaneously providing wetland habitats. The architects for the children's hospitals and the medical professionals had the courage to give the children freedom to use the natural spaces and trust the children and their carers to manage their play appropriately for their health conditions. This trust is most evident in Jeugdland Maakland, the adventure play area on the outskirts of Amsterdam, where children independently build structures using tools. These examples are just a taste of an increasing number of initiatives that demonstrate the benefits that children gain when nature is accessible and the benefits that come from thinking differently and ecologically about children's lives and the spaces they need and use.

City-level spaces: RioTinto Naturescape: Kings Park, Perth, Australia[6]

Built at a cost of AUS$10 million, the RioTinto Naturescape was designed as a destination natural play experience located in a bush landscape in a central city park. The 60,000 m^2 naturescape site includes a river with moveable boulders, log bridges, wetlands, climbing ropes, tree hides, water features, mud and many more features, all in a bush setting (Figure 11.5 and 11.6). This free park has won awards for its landscape design, but, more important, it has proved to be a massive success with Perth families. The naturescape is used by visiting families

Figure 11.5 Creating dens using the natural materials available at RioTinto Nature Space, Kings Park, Perth.

Figure 11.6 The river at RioTinto Nature Space, Kings Park, Perth; children play in the stream and the pond, move rocks and build dams.

mainly as a play space but has a more formal environmental education function when used by visiting groups. In the naturescape, 'nature' constitutes themed biological zones. These zones reference particular landscape types found within the Swan Coastal Plain, the geographic unit within which Perth is sited, and includes Paperbark Creek and a wetland (Bolleter, 2014). Perth's urban sprawl and almost messianic tidying up of natural areas means the naturescape presents a now rare opportunity for Perth's children to play in a natural landscape. However, the natural play offering is not without its more controversial side as Bolleter (2014) explains.

> The design also skilfully masks the contradiction that lies at the heart of such nature play areas, that their apparent informality and mutability are actually highly designed, regulated and very expensive. The regulated nature of such playgrounds is evident from the conditions of entry, which include no unaccompanied children, no swimming and no throwing rocks, among others. While nature play areas are typically designed for children twelve years and younger, it will be interesting to see how nature play environments can be adapted to the needs of teenagers, who are particularly sensitive to overt control or surveillance by adults.

Excellent though the naturescape park is, it is vitally important that it is not seen as a substitute for natural bush habitats. Children still need access to natural play spaces close to their homes.

Neighbourhood spaces: Bishan-Ang Mo Kio Park, Singapore

Nature can enhance children's lives and the places that children and their families use. The example of Bishan-Ang Mo Kio Park,[7] constructed in 1988, shows what can be done in a high-density environment. The park is traversed by the Kallang River, which was previously contained in a concrete canal. Given the neighbourhood's high-density high-rise housing, the park is a vitally important green space for families. In 2006, the Active, Beautiful, Clean Waters Programme (Public Utilities Board, 2016) was launched with the intention of rehabilitating rivers that had been canalised into more ecologically appropriate features to provide new recreation spaces for communities. The Kallang River was a flagship rehabilitation project. Families use the now more natural river to fish, particularly children fishing for guppies. The river now also benefits from the presence of a family of otters (Figure 11.7 and 11.8). Fishing is a popular activity for Singaporean families, with lakes experiencing high-density fishing, such as the fishing pond in Pasir Ris Park (see Chapter 5). In contrast to Pasir Ris, the rehabilitated Kallang River provides a much more natural fishing experience for children. The river also provides a safe and accessible natural water encounter, encouraged through the well-designed stepping-stones and natural bays along the river.

Figure 11.7 A rehabilitated part of the Kallang River in Bishan-Ang Mo Kio Park, Singapore.

Figure 11.8 Families out in Bishan-Ang Mo Kio Park, Singapore, showing more formal lawns on the left and more natural planting on the right along the tributary.

Jeugdland Maakland, Amsterdam[8]

Owned by the municipality of Amsterdam, Jeugdland describes itself as

> A green oasis where children can play outside the whole year. Such a 'natural playground' is of great importance in the city, with its lack of nature and confined playgrounds At Maakland, children are given the opportunity to develop in the fields of nature, culture, technology and science . . .[9]

Children have free entry to the site and to the workshops. On site, children can use tools to construct various buildings, play on site, or use the centre's facilities. Children can play freely without adult supervision. There are also children's vegetable gardens and a farm. Families can use the site free as well. On our weekday afternoon visit, children were using saws and tools and building structures from the wooden pallets, and families were visiting the city farm (Figure 11.9 and 11.10). Clearly this was a well-organised initiative but one where the children rather than the professionals designed the structures. Again, its success is in large part due to the recognition given to children's own competence to organise their play, use the tools and treat the various features, such as the pond, with care and respect.

Figure 11.9 Children working on one of the structures at Jeugdland Maakland.

Figure 11.10 The city farm forms part of the same complex and is open for families.

Biodiverse service spaces

Hospitals[10]

For children, hospitals can be stressful places to visit as both patients and visitors (Whitehouse et al., 2001). An evaluation of San Diego children's hospital's healing garden found that for most children, the garden was an important place to get away from the hospital but was underused because of accessibility issues, distance and lack of knowledge about the garden. The garden was important for visiting siblings but difficult for seriously ill children to use. Many hospitals are now reaching the end of their useful lives and being rebuilt. This rebuilding provides an excellent opportunity to incorporate the healing provided by green spaces. Bringing nature into the hospital as an integral component provides a better experience for visiting families and the seriously ill.

Two Australian children's hospitals take seriously the belief that nature heals. The Lady Cilento Hospital, Brisbane, which is Australia's largest children's hospital, was built at a cost of AUS$1.5 billion (2020 Vision, 2016). The hospital was designed around the belief that green design aids recovery and reduces stress for patients and their families. The hospital includes healing gardens and community spaces. It has eleven rooftop gardens and various communal spaces. The hospital recognises that quantifiable research outcomes would help future projects in justifying the green design by demonstrating positive impacts on heath expenditure. Another Australian children's hospital, the Royal Children's Hospital in Melbourne, built for AUS$1 billion, was designed as a hospital in a park (Healthy Parks Healthy People, 2017). The aim in designing the hospital was 'to show how healthcare spaces infused with nature can speak to children and aid in the healing process' (Healthy Parks Healthy People, 2017). In designing the building, the architects looked to the work of international thinkers such as Scandinavian artist Olafur Eliasson, psychologists Rachel Kaplan and Stephen Kaplan, and others on the relationship between nature and well-being. Nature was considered in every aspect as the lead architect, Kristen Whittle, explains,

> We took references of the eucalypts in the park and literally mapped them onto the building. So, the outside of the building is like a tree, it's built like tree bark and is made out of different layers of the grey colours of the trees.
> (Healthy Parks Healthy People, 2017)

Inside the building, they included a seven-by-four-meter aquarium and a meerkat enclosure. The hospital is built around a street that provides access to the park and has large picture windows that provide park views from the inside. Most children will have a view of the park from their rooms. Finally, Great Ormond Street, which is a busy central London children's hospital, recognises the healing effect of nature for children in a very specific way. The hospital has installed 'interactive digital wallpaper' along the corridors children pass through on their journey to theatre. Beneath the wallpaper are animated woodland creatures, such as deer, hedgehogs and birds that accompany children on their journey to theatre and the anaesthetic room. The walls distract children, reducing their stress and anxiety.

Schools[11]

After the home, the school is the environment where children spend most of their time. As many children's ability to free-play in the neighbourhood is constrained and as gardens decrease in size or are unavailable, schools can be the place where children have the most frequent opportunity to free-play in a green space. In Canada, it is estimated that children spend on average 110 minutes a day on school grounds for an average of 200 days per year. Thus, they are "worthy of attention" (Dyment and Bell, 2008, p. 952). The findings from this study were based on responses from 59 schools; they found that "green school grounds invite children to get moving in ways that nurture all aspects of their health and development" (Dyment and Bell, 2008, p. 960). Another study, also from Canada, compared children from schools with barren and biodiverse school grounds. It found that children's outdoor experience was richer among elementary school children attending biodiverse schools, their preferences were more nature oriented and the school ground offered more spaces, not just for play but for reflection and conversation (Samborski, 2010).

In light of the findings from these and other similar studies, there has been an international push to green school grounds. In Japan, Ikiminami Primary School won a gold prize at the Kids' Design awards for its project that transformed a courtyard/carpark into a school biotope garden. The garden has 52 species of plants, and the children have been shown to play in 180 different ways in the garden (Ito et al., 2010). This development also identified some issues related to creating such a garden; namely, it is a major time commitment, and it was difficult to retain continuity, as teachers in Japan are rotated between schools. However, the fact that teachers, students and the community have been directly engaged in the project means there was a high level of commitment to its success in the long term.

In the US, GreenSchoolyards America is a national organisation advocating green schoolyards. In the UK, Learning through Landscapes is a national charity advocating on behalf of and working with schools to create greener learning opportunities for schools. A survey they undertook of schools with grounds that had been greened revealed the following:

- 85% of teachers said it resulted in more creative learning and environmental awareness.
- 85% said that healthy active play had increased.
- 65% observed improved attitudes to learning.
- 73% said behaviour had improved.
- 64% reported reduced bullying.
- 84% reported improved social interaction

(Learning through Landscapes[12])

Also in the UK, Natural England has worked with children from 125 schools across the southwest[13] to enable them to experience learning in natural environments. Part of their remit is to encourage children and teachers to make use of natural spaces close to their school and within school grounds. They found 95% of children said outdoor learning makes lessons more enjoyable, 90% said they felt happier and healthier and 72% said they got on better with others (Natural England, 2016). In New Zealand, the Enviroschools movement "supports children and young people to plan, design and implement sustainability actions that are important to them and their communities" (Enviroschools, NZ, 2017). The movement has over 800 participating schools, and for these schools, green school grounds are an important element.

Greening school grounds takes several forms. For some schools, especially inner-city schools with minimal green space, it can be the introduction of natural elements to counteract the hard surfacing (Figure 11.11). One irony that exists is that sometimes schools have green spaces but declare these 'out of bounds' (Figure 11.12). Some schools have introduced ponds and others wildlife gardens to encourage invertebrates (Figure 11.13 and Figure 11.14). An excellent example of a school biodiversity project is that of St Malachy's School, which began their project in 1998, aided by Better Belfast, Conservation Volunteers and Ford. A Google search for the school shows the central city school as a green oasis in an otherwise heavily developed city setting[14] (Figure 11.15). Another innovation is

Figure 11.11 Introducing more natural features to a school playground in central London, UK.

Figure 11.12 Some schools that do have good open spaces, as in this photo, declare them out of bounds, preferring students to stay within the more formal play areas.

Figure 11.13 The 'bug' garden at a school in Auckland, New Zealand.

Figure 11.14 A wildlife pond at a school in Northern Ireland.

Figure 11.15 St Malachy's school, a green oasis in inner-city Belfast.
Source: Google Maps

the introduction of what are sometimes called 'loose parts', i.e., materials such as logs and brush that can be used to make climbing structures or dens. A recent and rapidly growing development is school gardens for growing vegetables.

Most important, schools should engage with the wider biodiversity present in their neighbourhood. This is especially important where schools are in high-density areas and have little on-site open space. Schools can adopt spaces nearby, such as leftover land, a stream or canal or part of a park. In our home city, Kaikorai Valley College[15] has helped in planting the streamside of Kaikorai stream with native plants. The school uses the stream for outdoor education and for studying water quality and flow, learning fly fishing and monitoring waste water. A recent project involved constructing more than 150 beehives in the school's technology workshops. Once schools start to engage with biodiversity projects and integrate sustainability into the school ethos and curriculum, other projects tend to follow. The school has used the regenerated stream for science projects and produced stream health data that has been valuable to the local council. Students as researchers is the next theme, where children are not just beneficiaries of biodiversity initiatives but contribute to building biodiversity knowledge.

Children as researchers and contributors to biodiversity knowledge

Citizen science is a growing area of research where non-professional volunteers contribute to authentic scientific research and conservation efforts. In a German study, school children aged 8 to 10 years were involved in a seed-counting experiment: the researchers found agreement in counts between those done by children and scientists, but children were less accurate on more sophisticated procedures (Miczajka et al., 2015, p. 1). The authors concluded that in citizen science projects, children could collect valuable scientific data as long as the task is at the right level for age and education and is relevant. There are many citizen science type projects designed to be suitable for children. Examples from the US, where children collect data that is used by university researchers include the following:[16]

- The Ant Picnic, where children use cookies to collect information about ants in their backyards (North Carolina State University)
- The Great Backyard Bird Count (Cornell Lab of Ornithology)
- The Ladybug Project, where children find, map and upload photographs of ladybugs (Cornell University)
- The Dark Sky Meter, where a phone app allows collection of data on light pollution

The range of projects is huge, and many help in developing better understandings of the biodiversity in children's own backyards. The following examples of citizen science projects aimed at children are taken from New Zealand. In these examples, children help scientists better understand cat predation on birds and provide information on moth species, locations and abundance.

Curious Minds: how safe is my cat?

Launched in 2014, Curious Minds[17] is a national government-funded initiative by the Ministry of Business, Innovation and Employment, the Ministry of

Education and the Office of Prime Minister's Chief Science Advisor. The programme works through representatives of science, education, business, the Māori, Crown Research Institutes, universities and other tertiary education institutions, as well as museums and various government agencies. The How Safe Is My Cat? project engaged around 70 primary school-aged children from four Dunedin schools, three rural and one urban, and their teachers, working together with the local ecosanctuary. The children's project was designed to answer the question of whether possum and rat traps set to catch pest species that are predators of native birds were a danger to their own domestic cats (Figure 11.16 and 11.17). New Zealand's wildlife developed in the absence of mammalian predators and lacks terrestrial mammals except for two species of bats. Cats are known to be significant predators of native wildlife, so while the project was ostensibly about keeping pet cats safe, its deeper goal was to educate children about the impacts their cats have on wildlife. To this end, children tracked their cats' movements using GPS collars, and they recorded visitors to baited but decommissioned traps using motion-sensitive trail cameras that were set up in the children's gardens and school grounds and recorded all the animals that visited the traps. The children developed hypotheses regarding what they expected would happen, mapped their GPS data to show where the cats went, viewed the motion sensor records identifying species that approached the traps and then analysed and presented the

Figure 11.16 Presenting findings from the Curious Minds cat project to an invited audience at Orokonui Ecosanctuary.

Figure 11.17 A poster produced on Reuben the cat, as part of the Curious Minds cat project.

data through posters and public events. The methods were scientifically robust. Research skills and communication were learnt, and, most important, the research findings can be used to inform pest eradication strategies being used to protect native birds in and around the ecosanctuary. The project has real-world scientific applications, but it also encouraged the children to think about predator-prey dynamics, determinants of cat behaviour, issues around invasive species and the biodiversity in their immediate environments. A related project has subsequently been developed working only in urban schools, with more than 50 families involved in collecting bird feeding, pest and predator data in their gardens. The school children are now very familiar with conservation tools such as predator traps, chew cards, tracking tunnels and movement-triggered cameras.

The Ahi Pepe MothNet[18] *– Aotearoa project*

This project, also located in Dunedin, is a citizen science project and a collaboration between a government research institute, Landcare Research, Orokonui Ecosanctuary, local Māori (Kati Huirapa Runaka ki Puketeraki) and university academics in Geography and Māori studies. Here, we focus on the contribution of one school, Te Kura Kaupapa Māori o Ōtepoti (a Māori language immersion school in Dunedin). Children worked with scientists to learn more about moths, and through moths gain a better understanding of wider ecology. Moths were trapped and then formally identified. Although each moth has a Latin name, the children gave the moths Māori names to help people see the moths as something familiar. The study deeply engages with Māori culture and scientific traditions. The data are helping scientists understand some of the 2,000 species of moths in

New Zealand, 90% of which are endemic to New Zealand, i.e., found nowhere else. The children presented their findings in Māori at a large national gathering of biodiversity scientists in the capital city, Wellington. The school then raised funds to send four children to the 2017 World Indigenous Peoples Conference on Education (WIPCE) in Toronto, Canada. In addition, each week, the children from Te Kura Kaupapa Māori o Ōtepoti present a show on local radio where they discuss science and interview a scientist. As with the cat project, the children are undertaking research using rigorous scientific methods and providing valuable data that can be used to enhance understanding of moths and moth ecology. Table 11.2 shows how these two projects stack up against the Department of Conservation's criteria for success.

Table 11.2 Analysis of the Curious Minds projects against the Department of Conservation's criteria for success

Department of Conservation criteria for successful projects	*Cat project*	*Ahi Pepe moth project*
Make it relevant to everyday life.	The children's own cats were involved, and the night cameras revealed the wildlife passing through their gardens and school grounds unseen at night.	Moths are present in all the children's environments.
Include families, communities and role models.	Families gave permission for the use of their cats and their gardens and came to the public presentations.	The project involved support from whanau (family) and the Māori community working with scientists.
Provide opportunities for social connections.	The children met child researchers from other local schools.	The project followed kaupapa Māori ways of learning together.
Promote direct experiences.	The children were involved in all aspects, from putting collars on cats to hypothesis testing and data analysis.	Children were involved in all aspects from putting out traps to collecting and presenting data on moth species.
Encourage free-choice learning.	Children took responsibility for different aspects of the project.	There were many different aspects to the project – art, radio, naming species – so children could engage in different aspects.
Foster the role of active stakeholder.	Children worked with communities and conservation organisations.	Children worked with communities, iwi and conservation organisations. The project validated the children's own language and Māori science. Children presented their findings at an international event, acting as ambassadors for the project.

Department of Conservation criteria for successful projects	*Cat project*	*Ahi Pepe moth project*
Target real local issues.	Native wildlife predation by cats is a problem in the children's neighbourhoods.	Very little is known about moths and their conservation status. Moths are a key species in local ecology.
Promote collective action.	The children were the catalysts for a community-based citizen science awareness-raising project.	The children were the catalysts for a community-based citizen science awareness-raising project and for working with the Māori community.

An assessment of progress

In a short chapter like this one, it is not possible to begin to do justice to the wealth of initiatives that are taking place globally. There has been a massive turnaround in consideration given to children by professionals in their day-to-day practice. At national and local levels, there has been a proliferation of strategies, policies and agreements relating to the need to consider children in development and to provide more sustainable living environments. There remains an urgent need to recognise children as relevant across the professions and across institutions, as, though there are some excellent projects, the experiences they offer tend to be small scale and selective rather than large scale and accessible to all children. Figure 11.18, 'The Collaboration Star', shows some of the key ways government, society and experts can contribute to creating better environments for children. For example, experts possess specialist knowledge on ecology that can be used in creating biodiverse school grounds and play areas and can help inform children on projects such as the citizen science ones. Within society, there are people who work with children in providing nature experiences. This includes families who are working in their own backyards and on greening projects and community members who are advocating for safe natural spaces for children to play locally and who are using knowledge of their children and their neighbourhoods to inform educational initiatives and local government plans. Governments, through local government plans, can set aside resources for neighbourhood greening projects and create policy guidance in which consideration of and consultation with children is mandatory. These three points of the star – government, experts and society – need to recognise and work in tandem with children, their families and the city environment in order to achieve biodiverse city environments for children.

Progress is being made in some professions. Professional bodies such as the Planning Institute of Australia (PIA) acknowledge that the profession has to recognise children's needs and perspectives, as in the need for "safe environments and conditions that nurture the development of children" (PIA, 2007).

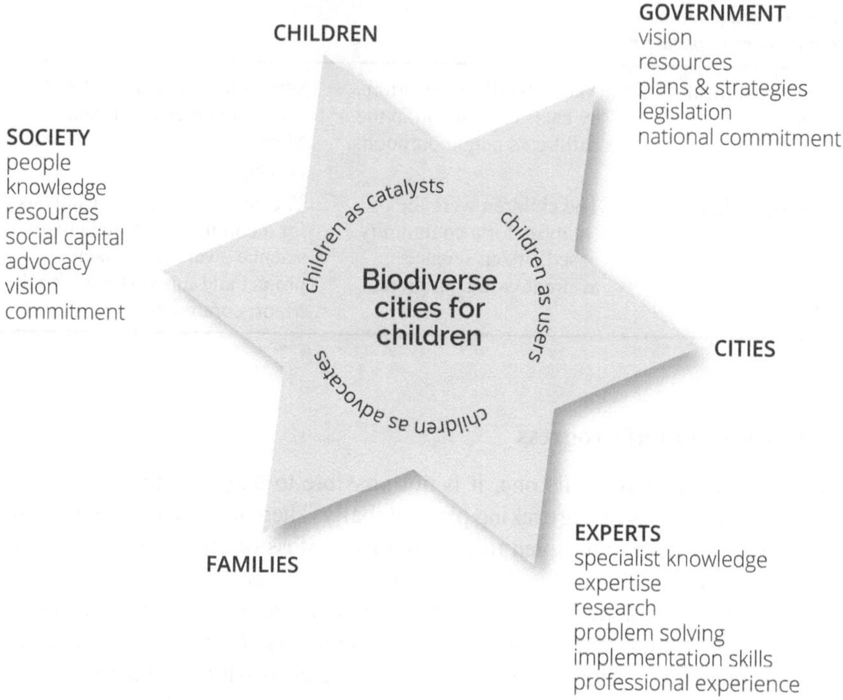

Figure 11.18 The Collaboration Star.

Nevertheless, they also recognise that there is "a growing tendency to design children out of built environments" and a "lack of research and dissemination of information to the built environment professions about children's issues". These statements have direct relevance to green space provision in Australia's built environment, yet practice still is moving inexorably towards less green space availability for children close to home, as garden size reduces and public space continues to be in competition with 'economic' space provision (Hall, 2010).

While initiatives such as those exemplified by the CTEET are inspiring, the reality is that life is challenging for many children in the city of Cape Town. The city grew by 45.9% between 1996 and 2011; 20% of the population lives in informal housing (a rate that is growing); 25% of the population is under 14; it has South Africa's highest crime rate with crime concentrated in poorer areas; the murder rate is very high at 51 per 100,000; and whilst some health trends are improving, child mortality, food insecurity and HIV rates are all high, and 31% still live below the poverty line (City of Cape Town, 2014). For Cape Town children, poor housing, their life experiences and movements will likely continue to be curtailed by poverty and fear of crime, and the pressures related to poverty mean nature connection will be low on the priority list. CTEET does a wonderful job, but even providing 30,000 child visits could be argued to be insignificant in a young and increasing population currently at 3.7 million, with approximately 1 million of these being children under age 15.

In less-challenging cities, concerted and committed action from professionals is for the most part still lacking. Most cities can point to some greening initiatives, but few can point to a coordinated plan to provide biodiversity across all children's environments, i.e., schools, services such as hospitals, public green spaces, private green spaces, biodiverse parks and streets. Few children can engage with wildlife and nature across all their life spaces, and few professionals are able to prioritise children over adults or biodiversity over economics in urban decision-making. Progress is being made, but it is still the exception. What progress there is takes place through a fragmented plethora of diverse policies and strategies that are often poorly used and understood, even by the professionals whose practice they are intended to guide. There remains a considerable gap between statements of commitment and practice around children, biodiversity and cities. Yet, as the Steel Valley Project in Stocksbridge, Sheffield (Steel Valley Project, 2017) shows, conceptualising an environment that works isn't difficult, and urban professionals could benefit their practice massively just by undertaking the project's Play Checklist as a central part of their development practice. The Steel Valley Project proposes a development where

> *Children play in gardens, parks, playgrounds, up trees, behind bushes, on walls, down banks, in puddles, outside shops, on the way to school . . .*

To see if planning developments are making a difference and a positive contribution to greener, more child-friendly cities, we end this chapter with the Steel Valley project's Play Checklist piece. This piece provides a great way to end a chapter but also a great place for individuals, parents and professionals to start by responding to the following questions:

Is there a space for my child to
run
jump
climb
crawl
roll
slide
balance
sit
hide
explore
shelter

dig
build
swing
cycle
skate
play ball
make dens
near the house?
If not, where could they do these activities?
Are these places close enough to the home?
Evidence shows that younger children tend to play within a distance of 100 m from their home.
Does my garden provide for a range of these activities?
Are these places safe and secure?
The busier a space, the safer it is for children's play.
Vehicles are always a threat to children.
Are the places they play in away from traffic?
If the answer is 'no', then are there ways in which the traffic could be slowed or reduced?
for example –
speed bumps
barriers and bollards
road narrowing
cul-de-sacs
dog-leg street layout
change in surfacing
closing streets to through traffic.
Is there a diversity of places near the home for my child to experience different situations –
small/large scale
enclosed/open spaces
hard/soft surfacing
busy/quiet?
Are there places for my child to interact with nature,
e.g., ponds, meadows, undeveloped 'wild' areas, wildlife gardens?
Are there quiet places away from people and traffic where my child can sit, talk, relax and enjoy being outside?
Are there larger areas where older children can experience more adventurous play?
Where green spaces seem too big for young children's play, use entrance spaces and create seating areas for adults.
Ultimately, children play anywhere and everywhere, so residential design and street improvements should always consider this. Children play in gardens, parks, playgrounds, up trees, behind bushes, on walls, down banks, in puddles, outside shops, on the way to school . . .

Notes

1 Housing has already been covered extensively in chapter 6.
2 Cape Town Environmental Trust: http://cteet.co.za/environmental-education-fund/.
3 The Cape Floral Region has been recognised as one of the most special places for plants in the world in terms of diversity, density and number of endemic species. The property is a highly distinctive phytogeographic unit which is regarded as one of the six Floral Kingdoms of the world and is by far the smallest and relatively the most diverse. It is recognised as one of the world's 'hottest hotspots' for its diversity of endemic and threatened plants and contains outstanding examples of significant ongoing ecological, biological and evolutionary processes, http://whc.unesco.org/en/list/1007.
4 Camley Street Natural Park: www.wildlondon.org.uk/reserves/camley-street-natural-park.
5 Bishan-Ang Mo Kio Park www.nparks.gov.sg/gardens-parks-and-nature/parks-and-nature-reserves/bishan-ang-mo-kio-park,
www.nparks.gov.sg/~/media/nparks-real-content/gardens-parks-and-nature/diy-walk/diy-walk-pdf-files/bishan-amk-park.pdf,
http://livebettermagazine.com/article/ch2m-hill-singapores-canals-reconnecting-people-to-waters-life-flow/.
6 RioTinto Nature Scape, Perth: www.healthyactivebydesign.com.au/case-studies/rio-tinto-naturescape,
www.bgpa.wa.gov.au/kings-park/area/naturescape/about-naturescape http://architectureau.com/articles/rio-tinto-naturescape/.
7 *American Society of Landscape Architects* 2016 ASLA Professional Awards, www.asla.org/2016awards/169669.html and www.straitstimes.com/multimedia/photos/park-reels-in-nature-lovers.
8 Jeugdland: https://maakland.nl/.
9 Jeugdland: https://maakland.nl/.
10 Lady Cilento Children's Hospital: http://202020vision.com.au/project/?id=356.
Melbourne Children's Hospital: www.hphpcentral.com/article/royal-childrens-hospital.
Great Ormond Street Hospital: www.gosh.nhs.uk/parents-and-visitors/gosh-arts/art-collection/nature-trail.
11 Ikiminami Primary, Japan: www.thenatureofcities.com/2013/07/21/growing-place-in-japan-creating-ecological-spaces-at-school-that-educate-and-engage-everyone/,
Green Schoolyards USA: www.greenschoolyards.org,
Enviroschools New Zealand: www.enviroschools.org.nz/.
12 Learning through Landscapes: www.ltl.org.uk/about/about-ltl.php.
13 Natural England: www.gov.uk/government/news/englands-largest-outdoor-learning-project-reveals-children-more-motivated-to-learn-when-outside.
14 St Malachy's School, Belfast: www.ourwildlifegarden.org/.
15 Kaikorai Valley College: www.odt.co.nz/news/dunedin/pupils-honey-huts-creating-buzz-nationwide.
16 Ant Picnic, The Great Backyard Bird Count, Dark Sky Meter: https://scistarter.com/project.
17 Curious Minds, Cat project: www.curiousminds.nz/about/curious-minds/.
18 Curious Minds, Moth project: www.landcareresearch.co.nz/information-for/citizen-science/shedding-light-on-the-night.

References

2020 Vision. 2016. *Designed for Healing: The Gardens of the Lady Cilento.* http://202020vision.com.au/project/?id=356

Bolleter, J. 2014. *Child's Play: Rio Tinto Naturescape*. http://architectureau.com/articles/rio-tinto-naturescape/Architecture Australia http://architectureau.com/articles/rio-tinto-naturescape

Cape Town Environmental Education Trust. http://cteet.co.za/ accessed 16–7–2017.

Child Poverty Action Group, London. 2017. www.cpag.org.uk/campaigns/child-poverty-london/keyfacts

City of Cape Town. 2014. State of Cape town 2014: Celebrating 20 years of democracy, City of Cape Town, Cape Town South Africa.

City of Freiburg. 2016. *Green City Freiburg, Approaches to Sustainability*, City of Freiburg im Breisgau, International Affairs Division 2016. www.freiburg.de/greencity

Department of Conservation. 2016. *New Zealand Biodiversity Action Plan 2016–202*, Department of conservation and the NZ Government, Wellington. www.doc.govt.nz/nature/biodiversity/nz-biodiversity-strategy-and-action-plan/ accessed 16–7–2017.

Dyment, J.E., and Bell, A.C. 2008. Grounds for movement: Green school grounds as sites for promoting physical activity. *Health Education Research* 23(6): 952–962.

Enviroschools, NZ. 2017. www.enviroschools.org.nz/about-enviroschools, accessed October 2017.

Hall, T. 2010. *The Life and Death of the Australian Backyard*. CSIRO Publishing, Collingwood, Australia.

Healthy Parks Healthy People. 2017. *Nature Heals: Melbourne's New Children's Hospital*. www.hphpcentral.com/article/royal-childrens-hospital accessed 16–7–2017.

International Making Cities Liveable. www.livablecities.org/articles/freiburg-city-vision accessed 16–7–2017.

Ito, K., Fjortoft, I., Manabe, T., Masuda, K., Kamada, M., and Fujiwara, K. 2010. Landscape Design and children's participation in a Japanese primary school: Planning process of school biotope for 5 years. In N. Müller, P. Werner, and J.G. Kelcey (Eds.), *Urban Biodiversity and Design*, Wiley-Blackwell, Oxford, UK, ch. 23.

London Wildlife Trust. 2017. *Camley Street Natural Park*. www.wildlondon.org.uk/reserves/camley-street-natural-park

Miczajka, V.L., Klein, A.-M., and Pufal, G. 2015. Elementary school children contribute to environmental research as citizen scientists. *PLoS One* 10(11): e0143229.

Natural England. 2016. www.gov.uk/government/news/englands-largest-outdoor-learning-project-reveals-children-more-motivated-to-learn-when-outside, accessed 16–7–2017.

Planning Institute Australia. 2007. *Child Friendly Communities*, PIA, ACT, Australia.

Public Utilities Board, Singapore. 2016. *Active, Beautiful, Clean Waters Programme*. www.pub.gov.sg/abcwaters

Samborski, S. 2010. Biodiverse or barren school grounds: Their effects on children. *Children, Youth, Environments* 20(2): 67–115.

Singapore Government National Parks. 2014. *Greening Schools for Biodiversity*. www.nparks.gov.sg/biodiversity/community-in-nature-initiative/greening-schools-for-biodiversity accessed 16–7–2017.

The Steel Valley Project Stocksbridge, Sheffield. 2017. www.thesteelvalleyproject.info/green/intro/childr-checklist.htm

Thorp, D. 2014. *The World's Most Successful Model for Sustainable Urban Development?* www.smartcitiesdive.com/ex/sustainablecitiescollective/words-most-successful-model-sustainable-urban-development/229316/ accessed 16–7–2017.

Whitehouse, S., Varni, J.W., Seid, M., Cooper-Marcus, C., Ensberg, M.J., Jacobs, J.R., and Mehlenbeck, R.S. 2001. Evaluating a children's hospital garden environment utilisation and consumer satisfaction. *Journal of Environmental Psychology* 21: 310–314.

Wilson, C. 2011. *Effective Approaches to Connect Children with Nature*, Department of Conservation, Wellington, New Zealand.

12 Towards better cities for children

What is nature? Nature is usually defined as everything that is not made by people, both living and non-living, and includes natural processes, such as wind, rain and earthquakes; in general, those things that have not been substantially altered by and persist despite human intervention. As described in Chapter 3, nature in urban environments is modified. Any species that thrives in a city must be tolerant of human disturbance and able to find sufficient resources to survive in a landscape that is dedicated to serving the needs of people (Figure 12.1). Cities are noisy, built up, paved over, lit up during the night, criss-crossed with roads and traffic and at all times inhabited by people. Yet nature in its modified form persists at multiple scales across diverse city landscape elements. The nature continuum we present in Chapter 1 seeks to represent the range of nature contexts that urban children could potentially experience.

While it would be wonderful if all children had safe access to truly wild places and the time and inclination to visit them, in this urbanising world, where the urban population is expected to surpass six billion by 2045, this is not an option (UN, 2014) (Figure 12.2a and 12.2b). In some countries, it is becoming increasingly difficult to find pristine, wild places anywhere; for example, just 12% of the United Kingdom is covered in woodland, and eight of 10 wooded areas are less than 20 ha in size (Spenser and Kirby, 1992), so that any organism living in a woodland is within 250 m of its edge. What most of today's and tomorrow's children will learn about nature and the benefits they will gain from engaging with nature will be the result of encounters that happen in cities. The ease with which children can access nature and the kind of nature available to them are going to depend very much on how we manage the small natural places. In this book we have argued that nature connections are not only desirable but achievable in nearly all urban contexts and that children can interact with nature in meaningful ways, even when nature is highly modified and at small scales.

An important reason for planning for enhanced opportunities for nature connection is improved well-being. Health benefits associated with nature contact in children are discussed in Chapter 7 and include improved cognitive functioning and attentional capacity and reduced stress and anxiety. Interventions in the form

Figure 12.1 These elvers (young eels) provide 'fishing' opportunities for this child with his sieve and bowl as they leave the harbour and migrate upstream through this concrete channel to more natural environments.

Figure 12.2a Though intended to be humorous, the message here is that these Australian children should not be allowed to have free, unsupervised play.

Figure 12.2b The freedom that these boys in Suva, Fiji, have is no longer available to many children growing up in western cities.

of various kinds of nature contact have been implemented for children of different ages to improve well-being, for example, greening of child-care facilities and school playgrounds to increase the number and quality of natural elements in order to stimulate more physical activity and increase imaginative play and the development of social relationships (Dadvand et al., 2015, Dowdell et al., 2011, Figure 12.3). In the case of youth at risk or youth with special needs, such as autism, nature-based therapies include farm therapy, such as horticulture and animal-assisted therapy (Annerstedt and Wahrborg, 2011; Haugan et al., 2006; Hine, 2008; Hine et al., 2008). Wilderness therapy comprises structured, nature-based activities in wild environments and has been used to address social and psychological problems in youth at risk of involvement in crime and children with ADHD (Annerstedt and Wahrborg, 2011; Levitt, 1994; Shanahan et al., 2009) (Figure 12.4). There are also structured programmes designed to facilitate children's play in natural environments to enhance their attention and learning, physical activity and mental health (Arbogast et al., 2009; Fjørtoft, 2001; Staempfli, 2008). For example, children in forest schools typically spend a proportion of their time at school outside engaged in outdoor activities (Eastwood and Mitchell, 2003; Murray, 2004; O'Brien, 2009; Rickson et al., 2004). These activities take place in a wide variety of environments which should include natural urban spaces.

Figure 12.3 At this school in Fiji, the grassy slope provides play opportunities for the youngest children as they slide down the bank; one is using a wooden board. Children have the capacity to find play value in a wide range of natural and semi-natural settings.

Figure 12.4 Watching two of the world's rarest penguins (yellow-eyed penguins or hoiho, *Megadyptes antipodes*) heading out to sea from a specially designated and protected breeding area. Such encounters are vital in developing wider conservation awareness and a necessary complement to more local urban experiences.

Definitions of nature in the literature devoted to well-being benefits have been very broad. What is not well known is whether the type of nature experience has an influence on the derived well-being benefits. Research driven by public health professionals has often used, from an ecologist's perspective, very simplistic representations of nature. Remote sensing is frequently the tool used in studies conducted on large numbers of people spread over large urban areas to quantify the amount of green present in people's neighbourhoods. But nature is more than just green. Is a large area of mown lawn, such as a sportsground, more valuable than a much smaller area of tangled woodland, perhaps with a small stream and a dense understorey, or a pond in a natural wetland setting, with its assemblage of aquatic and terrestrial vertebrates and invertebrates? Shanahan et al. (2015) emphasise the need to think beyond simplistic measures of 'nature dose' to better inform city planning and recommendations for minimum levels of exposure. They argue for guidelines on how much or how frequently people need to engage with nature and provide empirical evidence that visits of 30 minutes or more to green spaces during the course of a week could reduce the prevalence of depression and high blood pressure in the adult population by up to 7% and 9% respectively (Shanahan et al., 2016). Although they included a measure of intensity of nature experience (vegetation complexity) in their analyses, they did not find an association with this and any of the health outcomes measured. However, other studies on perceptions of diversity and psychological well-being have shown that benefits do increase with greater species richness of plants and habitat diversity (Fuller et al., 2007;

Dallimer et al., 2012) and bird abundance (Cox et al., 2017). Gross structural habitat heterogeneity is thought to act as a cue to biodiversity and associated benefits, in which case a mosaic of habitat patches may serve to enhance biodiversity, ecosystem services and human well-being (Fuller et al., 2007). However, there is considerable scope for improving our understanding of the relationship between biodiversity, habitat structure and well-being benefits.

All of the studies mentioned in the preceding paragraph have been carried out on adults; what do they mean for children? Do children perceive nature in their environments differently than adults? Drawings of nature by young children presented in Chapter 2 indicate that pets are commonly featured, and there is a focus on naming species, with little discrimination as to whether they are native or not; for example, some of the 6-year-old children interviewed for this book included poisonous frogs and eagles in discussions on nature, neither of which exist in New Zealand. Younger children (5 to 6 years) appear to see themselves as being part of nature rather than as a separate entity, affiliating with the 'integrated lens' (Figure 1.2) way of seeing nature. Older children (10 to 11 years) were more reflective of nature as an entity that encompasses multiple elements, and they are more likely to articulate emotions; nature as peaceful and vulnerable. Payne (2014) found that 11- to 12-year-old children's conceptions of nature were to view nature as something external to them and ideal, in which humans had no part. He discussed the apparent contradiction between these conceptions and everyday experiences with nature in their school grounds. What we do not know is how well children perceive diversity of species and habitats and whether perceptions affect the depth of nature connection and its associated benefits. However, it does appear that nature encounters experienced as a child are necessary to nurture biophilic attitudes towards wildlife (Zhang et al., 2014), leading to enhanced willingness to support conservation initiatives as adults. In particular, collecting insects and plants was a more important factor than socio-demographics in predicting positive attitudes towards wildlife as adults (Hosaka et al., 2017). In general, the more nature-based activities that people participate in during their childhood, the more they seek such activities when they are adults and overcome constraints to participation (Ward Thompson et al., 2008; Asah et al., 2012).

Recommendations made for promoting adult health by enhancing nature contact have direct consequences for children in terms of opportunities to connect with nature. For example, doses of nearby nature, such as would be found in private gardens, are particularly associated with quantifiable health benefits in adults, emphasising the value of this significant urban habitat. Since most children's independent movements are limited to within about 500 m of their homes with private gardens as preferred spaces (Hand et al., 2016, 2017) and since it is unlikely that truly wild spaces occur within this range of movement, gardens emerge as important biodiverse spaces in which children can encounter nature. Chapter 4 describes the many roles of gardens as highly valued spaces for food production, recreation, socialising and relaxation and as sites for activities such as bird feeding. Gardens have the potential to support significant biodiversity and ecosystem services and provide widespread connectivity for wildlife populations (Figure 12.5). In areas

Figure 12.5 Searching for caterpillars in the family vegetable patch.

of high-density living, the role of private gardens can be filled by nearby public green spaces and community gardens.

Given the evidence for well-being benefits, we have argued that access to nature should be a fundamental right but that it is necessary to develop a better understanding of how children see and interact with the natural world. It is all very well to identify the 'nature dose' needed and use information gathered on the type of nature experience that yields the best outcomes to guide the design of green spaces, but children have to be able to access these spaces and want to visit them, and clearly a number of barriers act to frustrate their opportunities to connect. Chapters 6 and 8 discuss the enormous role culture plays in influencing children's nature experiences, including the context of how the family relates to nature and also parents' beliefs that either facilitate or constrain their children's opportunities to interact with nature or play in natural spaces. Just having a pet can promote a more bio-centric life (Figure 12.6). A couple of caged birds, such as budgerigars, can cause a child to consider different aspects of bird behaviour, diet, song and moult, while walking the dog can take children to natural places they might not access otherwise. Family holidays often expose children to wild spaces with more natural play opportunities – rock pools at the beach, creatures under rocks in streams, mountains, lakes or forests. Parents also provide vicarious experiences of nature through the books they read to their

Figure 12.6 Pets repeatedly feature in the authors' research with children. Through pet ownership, children learn a sense of caring for animals and also about life cycles, as pets such as rabbits often have short lives.

children or the films they allow their children to see, although the representations of nature in books and film are typically of pristine spaces, almost always rural and, in the case of documentaries, focus on the fabulous or rare. Depictions of urban nature are uncommon and often associated with negative messages. As children grow older, the family lens is modified by the societal lens, such that children's perception of the world is determined by socio-economic, ethnic, religious, age and gender factors present in their family but also in their community and their nation.

Attitudes of parents towards nature are crucial in determining the way in which they introduce the natural world to their children. Adults themselves may have uneasy feelings towards neighbourhood green spaces which are transmitted to their children. Improvements in self-reported well-being with increasing naturalness in a green space sometimes can start to decline when the naturalness reaches a certain point and may become negative (Kaplan and Kaplan, 1989; Van den Berg and Heijne, 2005). These feelings are in response to perceptions of nature as scary, uncomfortable, dangerous and disgusting (Herzog and Kropscott, 2004; Andrews and Gatersleben, 2010; Sreetheran and van den Bosch, 2014). In some cities, parents' reluctance to allow their children access to green spaces is justified. Problems with antisocial elements, such as the homeless, hostile strangers, drug users, alcoholics, gangs and crime are common reasons for not using local parks (Byrne and Wolch, 2009; Cohen et al., 2007; Gobster, 2002). In Cape Town, reasons for non-use of community neighbourhood parks also included fear of racial attacks; the park being too crowded, too far away or too small; lacking facilities such as play equipment; and lacking in trees and nature (Willemse and Donaldson,

2012). Residents complained that parks were dry fields, often without any trees or grass, presenting a barren, dull space that was too small. Cape Town is an example of a challenging environment, where apartheid planning has led to poor provision of services in townships, including green spaces, for the black majority (Willemse and Donaldson, 2012). The absence of private garden spaces has meant community neighbourhood parks are particularly important for recreation and leisure and are well visited except by some ethnic groups (Figure 12.7). Almost half (48%) of children in Willemse and Donaldson's (2012) survey visited community neighbourhood parks one to three days a week, and 24% visited four to seven days a week, yet these do not comprise opportunities to connect with nature. Instead, programmes such as those provided by the Cape Town Environmental Educational Trust must fill that gap. Although Cape Town has many community neighbourhood parks, only 14% of them are located in low-income residential areas, where townships are located, and the ratio of a single available park to the number of people is high (1:7,102) compared to high-income areas (1:1,430). In smaller urban centres in South Africa, McConnachie and Shackleton (2010) reported that poor income suburbs have up to 14 times less public green space per capita than more affluent ones. Attempts to redress this environmental injustice are made more challenging as pressure to produce large-scale low-cost housing under the Reconstruction and Development Plan resulted in housing developments that have very little green

Figure 12.7 This street tree provides a good play and hanging out place for children in East London, South Africa.

space, the residents of which are least able to afford to travel and access spaces outside their suburbs or towns.

Socio-economic status is a significant driver of green space accessibility and use by families, even in wealthy, countries such as Australia, the UK and the US (summarised in Jones et al., 2009), although not necessarily in all cities (Jones et al., 2009). In their survey of residents in Bristol, UK, Jones et al. (2009) found that survey respondents from more deprived areas lived closer to green spaces but accessibility and safety were perceived as poor, resulting in less frequent use. In this case, social factors modified the relationship between what was provided and what was perceived as available. Perceptions of accessibility can be complex, with poor agreement between perceptions of accessibility of public parks and objective measurements of accessibility among residents in Glasgow (MacIntyre et al., 2008). In some situations, the quality of the green spaces could be lower, as was found in Melbourne parks (Crawford et al., 2008), which had fewer amenities to support physical activity in children; that might discourage use, but this is not always the case.

If the goal is to create spaces that are biodiverse, it is critical to apply our understanding of the features driving species distributions, abundances and assemblages in urban areas. Chapters 10 and 11 describe the variety of spaces most relevant to children in which nature could be enriched and provide examples of where this has been done well. There is plenty of ecological knowledge available to inform planning and restoration initiatives. Schoolyards, for example, are the focus of the organisation 'Green Schoolyards America'. Green schoolyards that provide natural areas enable children to escape stress and build competence and supportive social groups (Chawla et al., 2014; Bagot et al., 2015) but also foster environmental learning and promote biodiversity, although to achieve this, their design must go beyond the simple 'veggie' garden (Wake and Birdsall, 2016). The school in Figure 12.8 is an Enviroschool and has ongoing projects to green the school grounds, including creating a lizard habitat (Figure 12.8).

Many studies carried out in a range of cities show that targeted planting to increase vegetation volume and understorey can serve to provide food for all foraging guilds and improves connectivity across the city. Trees along median strips and verges will facilitate the movement of animals across roads, which otherwise can be significant barriers to movement. Larger patches of natural habitat are known to support more species (e.g., in Sydney, the number of ant, plant and bird species dropped off sharply in habitat fragments less than 4 ha in size; Drinnan, 2005); however, if it is not possible to set aside larger areas, we must focus on making the small areas as diverse and well-connected as possible. Inevitably, there will be species that children will not be able to encounter without travelling to appropriate habitat outside the city. Nevertheless, by planting to support native species, managing parks less intensively (i.e., less mowing, pruning and removing of dead trees), managing recreational green spaces such as golf courses to support biodiversity and designing buildings, public lighting, and transport corridors with the requirements of other species in mind, the form of the city can evolve to accommodate other species and, by doing so, become a place where nature connections are embedded in everyone's lives.

Figure 12.8 Greening their school ground, this school has a range of environmental projects (a) Vegetable garden (b) Natural materials and quiet spaces (c) Information posts on native species (d) The lizard garden.

d

Figure 12.8 (Continued)

Planning at this scale requires the cooperation of many individuals and agencies, as outlined in Chapter 11. The importance of the social dimension in determining the success of initiatives cannot be over-stated. As described in Chapter 4, human values, beliefs and preferences drive the nature of our cities from the bottom up and the top down. Human activities, such as bird feeding and wildlife gardening, reflect these values and provide people with a simple way to connect to nature in their immediate environments. Through planting, individuals and communities can be effective in supporting ecosystem services and mitigating some of the less welcome consequences of urbanisation through cooling, air filtration and reduction of the heat island effect. The message of this book is that cities need not be a nature-deficient environment, even though the nature found in urban green spaces may be highly modified. Because children are more accepting of modified nature as well as what is ecologically authentic, they can enjoy nature encounters in many kinds of small spaces. The challenge is to design cities to make these spaces as biodiverse and as accessible as possible.

References

Andrews, M., and Gatersleben, B. 2010. Variations in perceptions of danger, fear and preference in a simulated natural environment. *Journal of Environmental Psychology* 30: 473–481.

Annerstedt, M., and Wahrborg, P. 2011. Nature-assisted therapy: Systematic review of controlled and observational studies. *Scandinavian Journal of Public Health* 39(4): 371–388. DOI: 10.1177/1403494810396400. PubMed PMID: WOS:000290757500005.

Arbogast, K.L., Kane, B.C.P., Kirwan, J.L., and Hertel, B.R. 2009. Vegetation and outdoor recess time at elementary schools: What are the connections? *Journal of Environmental Psychology* 29(4): 450–456.

Asah, S.T., Bengston, D.N., and Westphal, L.M. 2012. The influence of childhood: Operational pathways to adulthood participation in nature-based activities. *Environment and Behaviour* 44(4): 545–569.

Bagot, K.L., Allen, F.C.L., and Toukhsati, S. 2015. Perceived restorativeness of children's school playground environments: Nature, playground features and play period experiences. *Journal of Environmental Psychology* 41: 1–9.

Byrne, J., and Wolch, J. 2009. Nature, race and parks: Past research and future directions for geographic research. *Progress in Human Geography* 33: 1–23.

Chawla, L., Keena, K., Pevec, I., and Stanley, E. 2014. Green schoolyards as havens from stress and resources for resilience in childhood and adolescence. *Health & Place* 28: 1–13.

Cohen, D.A., McKenzie, T.L., Sehgal, A., Williams, S., Golinelli, D., and Lurie, N. 2007. Contribution of public parks to physical activity. *American Journal of Public Health* 97(3): 509–514.

Cox, D.T.C., Shanahan, D.F., Hudson, H.L., Fuller, R.A., Anderson, K., Hancock, S., and Gaston, K.J. 2017. Doses of nearby nature simultaneously associated with multiple health benefits. *International Journal of Environmental Research and Public Health* 14: 172. DOI: 10.3390/ijerph14020172.

Crawford, D., Timperio, A., Giles-Corti, B. et al. 2008. Do features of public open spaces vary according to neighbourhood socio-economic status? *Health & Place* 14: 889–893.

Dadvand, P., Nieuwenhuijsen, M.J., Esnaola, M., Forns, J., Basagana, X., Alvarez-Pedrerol, M. et al. 2015. Green spaces and cognitive development in primary schoolchildren. *Proceedings of the National Academy of Sciences* 112(26): 7937–7942. DOI: 10.1073/pnas.1503402112. PubMed PMID: WOS:000357079400034.

Dallimer, M., Irvine, K.N., Skinner, A.M.J., Davies, Z.G., Rouquette, J.R., Maltby, L.L., Warren, P.H., Armsworth, P.R., and Gaston, K.J. 2012. Biodiversity and the feel-good factor: Understanding associations between self-reported human well-being and species richness. *BioScience* 62(1): 47–55.

Dowdell, K., Gray, T., and Malone, K. 2011. Australian journal of outdoor education: Nature and its influence on children's outdoor play. *Australian Journal of Outdoor Education* 15(2): 24–35.

Drinnan, I.N. 2005. The search for fragmentation thresholds in a Southern Sydney Suburb. *Biological Conservation* 124(3): 339–349.

Eastwood, G., and Mitchell, H. 2003. *An Evaluation of the First Three Years of the Oxfordshire Forest School Project*, Oxfordshire County Council, Oxford.

Fjørtoft, I. 2001. The natural environment as a playground for children: The impact of outdoor play activities in pre-primary school children. *Early Childhood Education Journal* 29(2):111–117.

Fuller, R.A., Irvine, K.N., Devine-Wright, P., Warren, P.H., and Gaston, K.J. 2007. Psychological benefits of green space increase with biodiversity. *Biological Letters* 3: 390–394.

Gobster, P.H. 2002. Managing urban parks for a racially and ethnically diverse clientele. *Leisure Sciences* 24(2): 143–159.

Hand, K.J., Freeman, C., Seddon, P.J., Recio, M.R., Stein, A., and van Heezik, Y. 2017. Children's biophilia: The importance of gardens in urban areas. *PNAS* 114(2): 274–279. www.pnas.org/cgi/doi/10.1073/pnas.1609588114

Hand, K.J., Freeman, C., Seddon, P.J., Stein, A., and van Heezik, Y. 2016. A novel method for fine-scale biodiversity assessment and prediction across diverse urban landscapes. *Landscape and Urban Planning* 151: 33–44.

Haugan, L., Nyland, R., Fjeldavli, E., Meistad, T., and Braastad, B.O. 2006. Green care in Norway – farms as a resource for the educational, health and social sector. In J. Hassink and M. VanDijk (Eds.), *Farming for Health,* Springer, Netherlands, pp. 109–126.

Herzog, T.R., and Kropscott, L.S. 2004. Legibility, mystery, and visual access as predictors of preference and perceived danger in forest setting without pathways. *Environment and Behavior* 36: 659–677.

Hine, R. 2008. Care farming: Bringing together agriculture and health. *Ecos: A Review of Conservation* 29(2): 42–51.

Hine, R., Peackock, J., and Pretty, J. 2008. Care farming in the UK: Contexts, benefits and links with therapeutic communities. *International Journal of Therapeutic Communities* 29(3): 245–260.

Hosaka, T., Sugimoto, K., and Numata, S. 2017. Childhood experience of nature influences the willingness to coexist with biodiversity in cities. *Palgrave Communications* 3: 17071. doi:10.1057/palcomms.2017.71

Jones, A., Hillsdon, M., and Coombes, E. 2009. Greenspace access, use, and physical activity: Understanding the effects of area deprivation. *Preventative Medicine* 49: 500–505.

Kaplan, R., and Kaplan, S. 1989. *Experience of Nature: A Psychological Perspective*. Cambridge: Cambridge University Press.

Levitt, L. 1994. What is the therapeutic value of camping for emotionally disturbed girls? *Women & Therapy* 15(3–4): 129–137.

MacIntyre, S., Macdonald, L., and Ellaway, A. 2008. Lack of agreement between measured and self-reported distance from public green parks in Glasgow, Scotland. *International Journal of Behavioral Nutrition and Physical Activity* 5: art no 26.

McConnachie, M., and Shackleton, C.M. 2010. Public green space inequality in small towns in South Africa. *Habitat International* 34: 244–248.

Murray, R. 2004. *Forest Schools Project Evaluation: A Study in Wales London.* New Economics Foundation, pp. 1–68.

O'Brien, L. 2009. Learning outdoors: The forest school approach. *Education* 37(1): 45–60.

Payne, P. 2014. Children's conceptions of nature. *Australian Journal of Environmental Education* 30(1): 68–75.

Rickson, M., Dillon, J., Teamey, K., Morris, M., Young Choi, M., Sanders, D. et al. 2004. *A Review of Research on Outdoor Learning*. National Foundation for Educational Research 148, UK.

Shanahan, D.F., Bush, R., Gaston, K.J., Lin, B.B., Dean, J., Barber, E., and Fuller, R.A. 2016. Health benefits from nature experiences depend on dose. *Scientific Reports* 6: 28551. DOI: 10.1038/srep28551

Shanahan, D.F., Fuller, R.A., Bush, R., Lin, B.B., and Gaston, K.J. 2015. The health benefits of urban nature: How much do we need? *BioScience* 65(5): 476–485.

Shanahan, L., McAllister, L., and Curtin, M. 2009. Wilderness adventure therapy and cognitive rehabilitation: Joining forces for youth with TBI. *Brain Injury* 23(13–14): 1054–1064.

Spenser, J.W., and Kirby, K.J. 1992. An inventory of ancient woodland for England and Wales. *Biological Conservation* 61: 77–93.

Sreetheran, M., and van den Bosch, C.C.K. 2014. A socio-ecological exploration of fear of crime in urban green spaces: A systematic review. *Urban Forestry and Urban Greening* 23: 1–18.

Staempfli, M.B. 2008. Reintroducing adventure into children's outdoor play environments. *Environment and Behaviour* 41: 268–280.

United Nations. 2014. *World's Population Increasingly Urban with More Than Half Living in Urban Areas*. www.un.org accessed 25–10–2017.

Van den Berg, A.E., and Heijne, M. 2005. Fear versus fascination: An exploration of emotional responses to natural threats. *Journal of Environmental Psychology* 25: 261–272.

Wake, S.J., and Birdsall, S. 2016. Can school gardens deepen children's connection to nature? In K. Nairn et al. (Eds.), *Space, Place and Environment, Geographies of Children and Young People 3*, Singapore: Springer Science+Business. DOI: 10.1007/978-981-287-044-5_1

Ward Thompson, C., Aspinall, P., and Montarzino, A. 2008. The childhood factor: Adult visits to green places and the significance of childhood experience. *Environment and Behavior* 40(1): 111–143.

Willemse, L., and Donaldson, R. 2012. Community neighbourhood park (COMMUNITY NEIGHBOURHOOD PARKS) use in Cape Town's townships. *Urban Forum* 23: 221–231.

Zhang, W., Goodale, E., and Chen, J. 2014. How contact with nature affects children's biophilia, biophobia and conservation attitude in China. *Biological Conservation* 177: 109–116.

Index

Note: Page numbers in italics indicate figures and those in bold indicate tables. Boxes and notes are indicated by page numbers followed by b and n, respectively.

For Product Safety Concerns and Information please contact our EU representative GPSR@taylorandfrancis.com Taylor & Francis Verlag GmbH, Kaufingerstraße 24, 80331 München, Germany